# SOLID STATE PHYSICS

# SOLID STATE PHYSICS

## Problems and Solutions

**László Mihály**
**Michael C. Martin**
*State University of New York*
*at Stony Brook*

**WILEY-VCH**

WILEY-VCH Verlag GmbH & Co. KGaA

All books published by Wiley-VCH are carefully produced.
Nevertheless, authors, editors, and publisher do not warrant the information
contained in these books, including this book, to be free of errors.
Readers are advised to keep in mind that statements, data, illustrations,
procedural details or other items may inadvertently be inaccurate.

**Library of Congress Card No.:**
Applied for

**British Library Cataloging-in-Publication Data:**
A catalogue record for this book is available from the British Library

**Bibliographic information published by**
**Die Deutsche Bibliothek**
Die Deutsche Bibliothek lists this publication in the Deutsche Nationalbibliografie;
detailed bibliographic data is available in the Internet at <http://dnb.ddb.de>.

© 1996 by John Wiley & Sons, Inc.
© 2004 WILEY-VCH Verlag GmbH & Co. KGaA, Weinheim

Printed in the Federal Republic of Germany
Printed on acid-free paper

**Printing**   Strauss GmbH, Mörlenbach
**Bookbinding**   J. Schäffer Buchbinderei GmbH, Grünstadt

**ISBN-13:** 978-0-471-15287-3
**ISBN-10:** 0-471-15287-0

# Contents

# 3   Electronic Band Structure

# 4   Density of States

# 5   Elementary Excitations

**PART II  SOLUTIONS TO PROBLEMS**

**1  Crystal Structures**                                                      **107**

## 5  Elementary Excitations                                      175

## 6  Thermodynamics of Noninteracting Quasiparticles             187

## 7  Transport Properties                                        203

# Preface

A number of years ago, one of us had the opportunity to attend lectures by a particularly talented and dedicated high-school science teacher, Miklós Vermes. The lectures were full of interesting and entertaining demonstrations, and he told the audience a lot about how the basic laws of physics fit together into a logical structure. Yet, when Professor Vermes was asked about what is the really important part in learning physics, he answered: "Solve homework problems. That way, you will get used to the stuff. In fact, 'getting used to it' is a major part of learning."

In this spirit, we hand this book to advanced undergraduate and introductory graduate students in Condensed Matter Physics. "Get used to it." When you are used to it, when you no longer stumble over every detail, you will be able to see the forest for the trees more clearly.

We wrote this book to satisfy another need as well: to help measure the progress of the students at midterms, finals, and comprehensive exams. Every professor must have a set of carefully guarded problems, appropriate to the final exam of the course she/he is teaching (better yet to have several sets, if you are teaching in several consecutive semesters). One would like to share and discuss these—presumably interesting—physics questions with the students, but practical considerations do not allow for this; if a problem is given out as a homework assignment, its value as a final exam problem is greatly and understandably reduced. Our goal here was to provide a volume of problems greater in number than the "critical mass" number which can be memorized. If a student remembers *all* of the solutions for a sufficiently large set of problems, he or she pretty much knows the subject (for proof, see the argument in the first paragraph).

The problems and solutions presented in this book stem from several years of teaching advanced undergraduate and introductory graduate solid state physics courses in the Physics Department of SUNY at Stony Brook. During these courses we used several of the excellent textbooks available; some of them are listed as references to the present collection of problems. Naturally, the problems were developed and organized each year, more or less in accord with the textbook used in that particular course. As we began to assemble the present collection, we planned to divide the problems into chapters corresponding to one of the standard organizations of the subject matter. However, we discovered that we do not need to be tied by the same constraints as the typical introductory textbook: A particular aspect of superconductivity may be effectively used to illustrate the concept of density of states; magnetism and charge density waves fit reasonably well under the umbrella of interacting electron systems.

We recommend the use of this collection in conjunction with one of the "standard" textbooks. The textbook will provide the backbone of organization very much needed for the first-time encounter with the subject. The instructor should pick the appropriate problems and assign them (in tandem with the problems from the textbook) as the course proceeds. Presumably, by the end of the course the problems *not* assigned can be used by the students for review and integrating their knowledge. We also hope that the problems solved here will provide inspiration for creating other, similar problems.

Although both authors' primary interests are experimental in nature, the reader will find a few problems concerning model systems with little practical relevance. In addition, there were several times when we felt that a particular concept is better illustrated if the mathematics remains simple, so we utilized one- or two-dimensional systems (which are much easier to make graphical illustrations of too!). We hope that we do not draw too much criticism for not being realistic or for neglecting important practical experimental issues. The more abstract problems in this collection may help the community of condensed matter physicists to preserve a common language so that an experimenter can still attend a theory talk without losing track within the first two minutes. Nevertheless, we tried to keep this book at an introductory level, and the problems using second quantization formalism do not really require more than the knowledge of the basic commutation relations.

Many people helped us eliminate some (but, we are afraid, not all) conceptual and practical mistakes in this book. We are particularly indebted to Jenő Sólyom, Manuel Cardona, György Kriza, Attila Virosztek, and Gábor Oszlányi for helpful comments. We would also like to thank Vladimir Golovanov for a critical reading of our manuscript. We would appreciate comments from readers (you can find us on the Internet* or contact us via the publisher).

Most of the graphs in this book were produced using CoPlot.[1] The typesetting was done with LaTeX and using the emTeX program.[2]

László Mihály<br>Michael C. Martin

*Stony Brook*
*May 1996*

[1]"CoPlot," computer plotting software, © 1988 and 1990 by CoHort Software, Berkeley, California.
[2]emTeX, TeX processing software, written by Eberhard Mattes and available free of charge on the Internet.
*at http://buckminster.physics.sunysb.edu/book/book.html

# PART I
# Problems

# 1 Crystal Structures

A *crystal* is a periodic array of atoms. Many elements and quite a few compounds are crystalline at low enough temperatures, and many of the solid materials in our everyday life (like wood, plastics and glasses) are not crystalline. Nevertheless, typical solids state physics texts start with the discussion of crystals for a good reason: The treatment of a large number of atoms is immensely simplified if they are arranged into a periodic order.

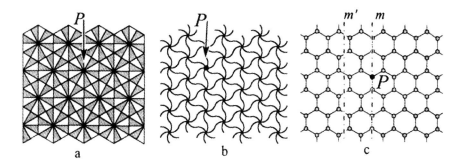

**Fig. I.1.1.** Examples of crystals in two dimension. Dots, curved lines or shaded areas represent various molecules or atomic arrangements.

Figure I.1.1 shows a few two-dimensional "crystals".[1] All crystals have *discrete translational symmetry*. If displaced by a properly selected *lattice vector R*, every atom moves to the position of an identical atom in the crystal. Due to this translational symmetry, a crystal can be constructed by repeating the *basis* at every *Bravais lattice* point. The basis is the "building block" of the crystal. It may be simple, a spherical atom, or as complex as a DNA molecule. Sometimes we have to use a basis made up of two (or more) atoms,

---

[1] Many textbooks, including Kittel [2], Ashcroft and Mermin [1], and Yu and Cardona [6] provide introductions to crystal symmetries. We refer the reader to those books for descriptions of three-dimensional crystals. Here we will use two-dimensional examples to illustrate the concepts.

3

even if there is only one type of atom in the crystal (see the example in Figure I.1.2).

The *Bravais lattice*, or *space lattice*, is an infinite array of points, determined by the lattice vectors $\boldsymbol{R}$, where $\boldsymbol{R} = n_1\boldsymbol{a}_1 + n_2\boldsymbol{a}_2 + n_3\boldsymbol{a}_3$ such that every $n_i$ is an integer. The $\boldsymbol{a}$'s are the three *primitive vectors* of the Bravais lattice; in three dimensions they must have a nonzero $\boldsymbol{a}_1 \cdot (\boldsymbol{a}_2 \times \boldsymbol{a}_3)$ product.

There are an infinite number of different choices for the primitive vectors of a given lattice. For example, $\boldsymbol{a}_1' = \boldsymbol{a}_1 + \boldsymbol{a}_2$; $\boldsymbol{a}_2' = \boldsymbol{a}_1 - \boldsymbol{a}_2$; and $\boldsymbol{a}_3' = \boldsymbol{a}_3$ will describe the same lattice. The *lattice spacings* are the lengths of the shortest possible set of primitive vectors.

All three crystals in Figure I.1.1 have the same Bravais lattice. Note that not all symmetric arrays of points are Bravais lattices! For example, Figure I.1.2 shows a *honeycomb lattice* and a choice for its Bravais lattice and basis.

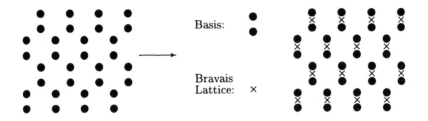

**Fig. I.1.2.** Example of a regular array of points that is not a Bravais lattice (honeycomb lattice).

In addition to the translational symmetry, most crystals also have other symmetries, including reflection, rotation, or inversion symmetry, or more complicated symmetry operations, like the combination of rotation and translation by a fraction of the lattice vector. In Figure I.1.1 the three-fold rotational symmetry around the point $P$ is common to all three crystals. The honeycomb lattice (Figure I.1.1c) also has a "mirror line" $m$, while the other two crystals do not have this symmetry. A less trivial symmetry operation is mirroring the honeycomb lattice with respect of the line $m'$, and then shifting it parallel to the line, until it overlaps with itself.

The collection of symmetry operations forms a *symmetry group*. The important property that defines a symmetry group is the relationship between the symmetry elements – i.e., what happens if two symmetry operations are applied subsequently. In the language of group theory, this relationship is described by the multiplication (direct product) table.[2] The symmetry group can be represented in many ways (collections of matrices, symmetry opera-

---

[2] An introduction to basic group theory is given by Yu and Cardona [6] pp. 21–43 or Harrison [5] pp. 16–20.

tions of a simple geometric object, and so on). As long as the multiplication table is the same, we are dealing with the same group. The crystals in Figures I.1.1a and I.1.1b have equivalent symmetry groups, while some of the symmetries of the honeycomb lattice are different.

When all possible symmetry operations are taken into account we talk about *crystallographic space groups*. Any given three-dimensional crystal belongs to one of the 230 possible crystallographic space groups. (Two-dimensional crystals are much simpler; there are only 17 inequivalent "crystallographic plane groups".) The symmetries are often identified by the name of a representative material, like "sodium chloride structure", "diamond structure", "wurzite (or zincblende, zinc sulfide) structure", and so on. More sophisticated group theoretical notations are used by crystallographers.[3] For a complex structure the identification of the symmetry group may be a rather nontrivial task.

A subset of symmetry operations that leaves at least one point invariant makes up the *crystallographic point group*. There are 32 different crystallographic point groups in three dimensions, and 10 in two dimensions. Considering the examples in Figure I.1.1, the rotations around $P$ and the mirror line $m$ are point group symmetries, but the combination of mirroring around $m'$ and the subsequent shift is *not* a point group operation.

Sorting out the symmetries of the Bravais lattices is much simpler. There are 14 different space groups for three-dimensional Bravais lattices, including the simple cubic (*sc*), face centered cubic (*fcc*), body centered cubic (*bcc*), simple tetragonal, body centered tetragonal, and others. Figure I.1.3 shows all possible Bravais lattices in two dimensions. It is important to emphasize that the symmetries of the Bravais lattice are intimately related to the symmetries of the original lattice. For example, the three-fold rotational symmetry of the honeycomb lattice results in the requirement that its Bravais lattice must have three-fold rotational invariance (which leaves the hexagonal lattice as the only choice, see Figure I.1.2).

Finally, when the point group symmetries of the Bravais lattices are considered, the choices are further limited, and in three dimensions only seven distinct groups are left. These define the seven *crystal systems*: Cubic, tetragonal, orthorhombic, monoclinic, triclinic, trigonal, and hexagonal. (In two dimensions there are four crystal systems. The rectangular and centered rectangular Bravais lattices shown in Figure I.1.3 make up the "rectangular" system.)

The *primitive unit cell* or *primitive cell* is a volume which will fill space completely, without overlap, if shifted by each of the lattice vectors. The primitive unit cell contains exactly one Bravais lattice point and the atoms in it can be used as the basis to construct the crystal. The volume made up by the primitive vectors is a possible primitive unit cell, but there are

---

[3] See Ashcroft and Mermin [1] pp. 122–126.

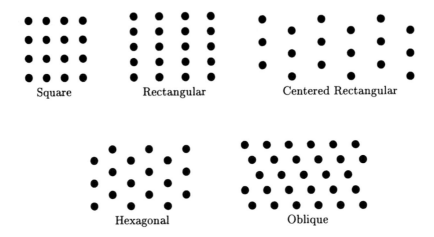

Fig. I.1.3 Two-dimensional Bravais lattices.

many other possibilities. More often than not, the primitive unit cell is less symmetric than the Bravais lattice.

The *unit cell* is a volume that fills up space with an integer multiplicity, if shifted by each of the lattice vectors. It contains an integer number of lattice points. Sometimes it is more convenient than the primitive unit cell (as shown in Figure I.1.4).

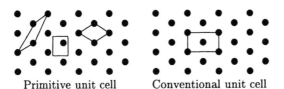

Fig. I.1.4. Some choices of primitive unit cells and the conventional unit cell for a centered rectangular lattice.

The *Wigner-Seitz cell* (WS cell) is a volume made up of space which is closer to a given lattice point than to any other point. There is a practical recipe for the construction of this cell: Select a lattice point, draw the lines connecting it to its neighbors (nearest and next-nearest is usually sufficient), and draw the perpendicular bisecting planes to these lines. The smallest volume enclosed within these planes is the WS cell. The WS cell is a primitive unit cell that preserves the symmetries of the Bravais lattice.

Imagine that a crystal is made of spheres with their diameters being equal to the nearest-neighbor distance. The *filling factor* is the volume fraction of the spheres relative to the total volume. The *coordination number* is the number of nearest-neighbors to any sphere.

When waves are scattered from a periodic array the constructive interference is often described by the *Bragg condition*: $n\lambda = 2d\sin\theta$, where $\lambda$ is the wavelength, $d$ is the spacing between subsequent lattice planes (that is, planes containing a high density of lattice points) and $\theta$ is the angle between the incident beam and the lattice planes. The scattered beam will have the same angle with respect to the planes as the incident beam, so the total scattering angle is $2\theta$.

Instead of wavelength, the concept of the *wavevector* is often used to characterize the a plane wave. The wavevector $k$ points in the direction of the propagation of the wave, and the magnitude of the vector is $|k| = 2\pi/\lambda$. The condition for constructive interference can be expressed in terms of wave vectors as $(k - k') \cdot d = 2\pi n$, where $n$ is an integer, $k$ and $k'$ are the incident and scattered wavevectors, and $d$ is the vector pointing from one scattering center to another.

The *reciprocal lattice* is a very useful tool to handle the diffraction of waves; it is generally used to describe all things of "wavy nature" (like electrons and lattice vibrations). Definitions of the reciprocal lattice are as follows:

- The collection of all wave vectors that yield plane waves with a period of the Bravais lattice (*note:* any $R$ vector is a possible period of the Bravais lattice).
- A collection of vectors $G$ satisfying $G \cdot R = 2\pi n$, or $e^{iG \cdot R} = 1$.
- There is also a practical definition in three dimensions:

$$g_1 = 2\pi \frac{a_2 \times a_3}{V} , \qquad (I.1.1)$$

and cyclic permutations of 1 2 3, where $V = a_1 \cdot (a_2 \times a_3)$ is the volume of the unit cell.

The volume of the primitive unit cell in the reciprocal lattice is $(2\pi)^3/V$. The crystal system of the reciprocal lattice is the same as the direct lattice (for example, cubic remains cubic), but the Bravais lattice may be different (*e.g.*, *fcc* becomes *bcc*). The *Brillouin zone* is the WS cell in the reciprocal lattice.

Using the reciprocal lattice, the condition for constructive interference becomes quite simple: If the difference between the incident ($k$) and scattered ($k'$) wave vectors is equal to a reciprocal lattice vector, the diffracted intensity may be nonzero. This is the *Laue condition*. With $K = k' - k$, this leads to the simple equation $K = G$, or $K \cdot G = 1/2|G|^2$. The *Ewald construction* is a geometric representation of these equations.

The *Miller indices*, $h$, $k$, and $l$, are obtained from the "coordinates" of a reciprocal lattice vector $G = hg_1 + kg_2 + lg_3$. By definition, the Miller indices are integers. For a simple cubic lattice these numbers are real coordinates in a Cartesian coordinate system.

There is an interesting relationship between Miller indices and lattice planes. For any plane there is an infinite number of other, parallel lattice planes, separated by a distance $d$. It is easy to see that the ratio $x : y : z$ is the same for all parallel planes, where $x$, $y$, and $z$ are the intercepts of a given plane with the coordinate axis defined by the primitive vectors $a_1$, $a_2$, $a_3$.

Sometimes the need arises to classify these planes. There is a convenient mapping between a given class of lattice planes and a lattice vector in reciprocal space: For any family of lattice planes separated by a distance $d$, there is a reciprocal vector with length $|G| = 2\pi/d$, and this vector is perpendicular to the lattice planes. One can show (nontrivially) that $h\colon k\colon l = (1/x)\colon(1/y)\colon(1/z)$.

The Laue condition is based solely on the Bravais lattice, so the positions of the diffraction peaks are independent of the atomic basis. However, the intensities of the peaks are strongly influenced by the basis. The *structure factor*, $S(G)$, and the *form factor*, $f_\alpha$, tell us how the intensities of the peaks depend on the atoms making up the crystals. These quantities are calculated as a sum (or integral) within the unit cell; therefore they may be totally different for two different crystals, even if the crystals have the same Bravais lattice. In the simplest approximation the scattering depends on the atomic charge distribution $\rho_\alpha(r)$, and the intensity is proportional to the absolute value squared of

$$S(G) = \sum_\alpha f_\alpha e^{-iG\cdot r_\alpha} \qquad (\text{I.1.2})$$

and

$$f_\alpha(G) = \frac{1}{e} \int \rho_\alpha(r) e^{-iG\cdot r} \mathrm{d}^3 r \;, \qquad (\text{I.1.3})$$

where $e$ is the electron charge, the sum is over the atoms in the unit cell, and the integration is over the volume of an atom. Similar formulae work for electron and neutron scattering, except the form factor integral is different depending on the microscopic interaction at play. Even for X-rays, the calculation of the form factor as an integral over the charge density works only for the simplest cases. For a realistic calculation of scattered radiation intensities one has to include factors representing the directional dependence of the scattering by a single point charge, the absorption of the radiation, and other effects. For powder samples this process is called *Rietveld analysis*. The expression becomes much more complicated for example, if there is a match between the energy of atomic transitions and the X-ray quanta.

When the atomic positions are time-dependent (for example, if lattice waves are excited in the crystal), the crystal scatters radiation at a frequency different from the incident frequency. In this case energy is either absorbed or emitted by the crystal. The process can be described by the *dynamic structure factor*, which depends also on the frequency difference $\omega$ between the incident and scattered radiation: $S = S(k, \omega)$. The general expression

$$S(k, \omega) = \frac{1}{N} \int \frac{\mathrm{d}t}{s\pi} e^{i\omega t} \langle \rho(k, 0)\rho(-k, t) \rangle \;, \qquad (\text{I.1.4})$$

relates the structure factor to $\langle \rho(k, 0)\rho(-k, t) \rangle$, the density–density correlation function. (Here $N$ is the number of primitive unit cells and $\rho(k, t)$ is the charge density at time $t$.) This formula is equally useful when dynamics of the system are described by quantum mechanics (and the $\langle\ \rangle$ expectation value is that of the density operators) or at finite temperature [when the (classical) atoms have thermal motion]. For a static array of classical atoms the quantity $\int d\omega S(\mathbf{G}, \omega) = S(\mathbf{G})$ is identical to the structure factor defined in Eq. I.1.2.

## 1.1 Problem: Symmetries

In Figure I.1.5, two "crystals" (a and b) and a polygon (c) are shown. Identify the point group symmetry operations of the three objects (assume that the crystals are of infinite size). Show that the point groups of the two crystals are different, and that one of them has the equivalent point group as the polygon.

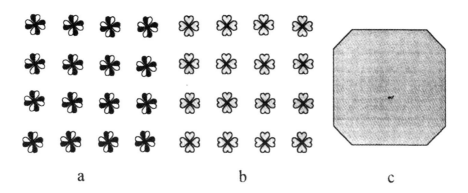

a                                  b                                  c

**Fig. I.1.5** Two-dimensional objects with four-fold rotational symmetry.

## 1.2 Problem: Rotations

A group is represented by three matrices:

$$\mathsf{E} = \begin{bmatrix} 1 & 0 \\ 0 & 1 \end{bmatrix} \qquad \mathsf{A} = \begin{bmatrix} \alpha & \beta \\ -\beta & \alpha \end{bmatrix} \qquad \mathsf{B} = \begin{bmatrix} \alpha & -\beta \\ \beta & \alpha \end{bmatrix}, \qquad (\text{I.1.5})$$

where $\alpha = \sin 30°$ and $\beta = \cos 30°$. Determine the multiplication table for this group. What is an example of a (two-dimensional) crystal with these point group symmetries?

## 1.3 Problem: Copper Oxide Layers

The common building blocks for most high temperature (high $T_c$) super-conductors are copper oxide layers, as depicted in Figure I.1.6. Assume the distance between copper atoms (filled circles) is $a$. For simplicity let us also assume that in the third dimension these $CuO_2$ layers are simply stacked with spacing $c$, and there are no other atoms in the crystal. In first approximation the layers have a four-fold symmetry; the crystal is tetragonal.

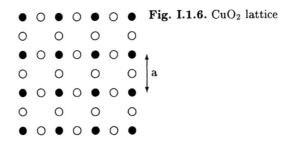

Fig. I.1.6. $CuO_2$ lattice

(a) Sketch the Bravais lattice and indicate a possible set of primitive vectors for this crystal. What is the unit cell, and what is the basis?

(b) In $LaCuO_4$ one discovers, at closer inspection, that the $CuO_2$ lattice is actually not flat, but that the oxygen atoms are moved a small amount out of the plane ("up" or "down") in an alternating fashion (in Figure I.1.7, a + means up and a - means down).[4] What is the primitive cell and lattice spacing for this crystal? What is the reciprocal lattice? Describe (qualitatively) what happens in the X-ray diffraction pattern as the distortion is decreased gradually to zero.

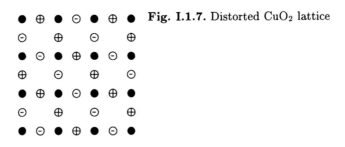

Fig. I.1.7. Distorted $CuO_2$ lattice

---

[4] $LaCuO_4$ is an antiferromagnetic insulator. High temperature superconductivity was discovered in a closely related compound, $La_{1-x}Ba_xCuO_4$. See J.G. Bednorz and K.A. Müller, Z. Physik B **64**, 189 (1986).

## 1.4 Problem: Graphite

How many atoms are in the primitive unit cell of graphite?

## 1.5 Problem: Structure of $A_x C_{60}$

Fleming and co-workers describe the structure of various alkaline metal-$C_{60}$ compounds in Fleming *et al.*, *Nature*, **352**, 701 (1991). Read the paper and answer the following questions:

(a) In Figure 2a of the paper an *fcc* structure of $A_3 C_{60}$ is indicated. In the usual representation of the *fcc* structure the $C_{60}$ molecules would be placed on the corners, *e.g.* at positions $(1, 0, 0), (0, 1, 0) \ldots (1, 1, 1)$, and the face centers $(\frac{1}{2}, \frac{1}{2}, 0), (\frac{1}{2}, \frac{1}{2}, 1) \ldots (0, \frac{1}{2}, \frac{1}{2})$ of a unit cube. The Figure certainly does not look like a cube. What is a possible choice for the "conventional" $(x, y, z)$ coordinates of the $C_{60}$ molecule in the middle of Figure 2a from this paper?

(b) Assume you perform a powder X-ray diffraction measurement on a Rb doped $C_{60}$ material with $\lambda = 0.9$ Å X-rays. You want to compare your results to Fleming's to see what stoichiometry your compound is. What are the positions ($2\theta$, in degrees) of the first five diffraction peaks for the three observed structures (doping $= 3, 4$, and $6$ in Table 2 of the paper)?

## 1.6 Problem: *hcp* and *fcc* Structures

$\alpha$-Co has an *hcp* structure with lattice spacings of $a = 2.51$ Å and $c = 4.07$ Å;[5] $\beta$-Co is *fcc*, with a cubic lattice spacing of 3.55 Å. What is the difference in density between the two forms?

## 1.7 Problem: *hcp* and *bcc* Structures

Sodium transforms from *bcc* to *hcp* at about $T = 23$ K. Assuming that the density remains fixed, and the $c/a$ ratio is ideal, calculate the *hcp* lattice spacing $a$ given that the cubic lattice spacing $a' = 4.23$ Å in the cubic phase.

## 1.8 Problem: Structure Factor of $A_x C_{60}$

Experiments show that the (2 0 0) X-ray diffraction peak of the *fcc* $C_{60}$ solid (lattice spacing: $a = 14.11$ Å) is very weak. Assume that the charge

---

[5] Descriptions of the *hcp* structure are given by Harrison [5] pp. 2–4, Ashcroft and Mermin [1] pp. 76–79, Ibach and Lüth [4] p. 23, and Kittel [2] pp. 23–24.

distribution of a fullerene ball is represented by a surface charge on the surface of a sphere of radius 3.5 Å. Calculate the form factor of the $C_{60}$ molecule in this approximation. Show that the form factor of the (2 0 0) peak is indeed much less than that of the (1 1 1) peak!

## 1.9 Problem: Neutron Diffraction Device

A collimated beam of neutrons of all energies is incident on a powder of cubic, monoatomic crystallites. What will happen to the neutron beam and what will the energies of the collimated neutrons exiting the powder be? How can this be used experimentally?

## 1.10 Problem: Linear Array of Emitters, Finite Size Effects

Consider a linear array of $N$ identical radiation sources, placed at positions $x_n = na$. Each source emits radiation at frequency $\omega$, wavenumber $k$, amplitude $A$ and the sources are coherent with zero phase difference. We detect the radiation with intensity $I_{detector}$ (defined as the time-averaged amplitude square) with a detector positioned along the line of the array, far away from the sources.

Assume $N \gg 1$, but finite. The $I(k)$ function will exhibit peaks of finite width. What are the main peak positions? What will the $I(k)$ function look like in the neighborhood of a main peak? What is the intensity of the detected radiation at the peak position? Estimate the width of this peak. How will the integrated intensity (integrated over $k$) depend on $N$?

## 1.11 Problem: Linear Array of Emitters, Superlattice

Consider a linear array of $N$ identical radiation sources, placed at positions $x_n = na$. Each source emits radiation at frequency $\omega$, wavenumber $k$, amplitude $A$ and the sources are coherent with zero phase difference. We detect the radiation with intensity $I_{detector}$ (defined as the time-averaged amplitude square) with a detector positioned along the line of the array, far away from the sources.

Introduce a static modulation to the position of the sources, described by $x_n = na + u_n$, where $u_n = u_0 \cos(qan)$ with $q < 2\pi/a$ and $u_0 \ll a$. In qualitative terms, what will we see in $I(k)$ as $u_0$ approaches zero? Discuss the peaks between $k = 2\pi/a$ and $k = 4\pi/a$ in greater detail: Calculate the first correction to the magnitude of the main peaks, and the magnitude of the strongest side peaks. (*Note*: Assume $N \to \infty$. The calculation of intensity becomes simplified if you concentrate on terms of order $N^2$.)

## 1.12 Problem: Powder Diffraction of *hcp* and *fcc* Crystals

Cobalt has two forms:[6] $\alpha$-Co, with *hcp* structure (lattice spacing of $a = 2.51$ Å) and $\beta$-Co, with *fcc* structure (lattice spacing of $a_{cubic} = 3.55$ Å). Assume that the *hcp* structure has an ideal $c/a$ ratio. Calculate and compare the position of the first five X-ray powder diffraction peaks. The quantity $K = 4\pi/\lambda \sin\theta$ can be used to characterize the peak positions (here $\lambda$ is the wavelength of the X-ray radiation and $2\theta$ is the scattering angle).

## 1.13 Problem: Momentum Resolution

A neutron source provides a monochromatic beam neutrons, with wavelength of $\lambda = 2.502 \pm 0.002$Å The divergence of the beam is $\delta\phi = 40$arc seconds. There is a collimator on the detector side, providing the same angular resolution. The sample has simple cubic structure of approximate lattice spacing $a = 3.55$Å.

Describe the experimental geometry for the most accurate measurement of the lattice spacing. What are the angles of the incident and diffracted beam relative to the crystal planes? Make a sketch of this geometry.

## 1.14 Problem: Finite Size Effects

Consider a finite size, simple cubic crystal of $N_x \times N_y \times N_z$ atoms, with lattice spacing $a$.

(a) For $k_x = k_y = 0$, plot the $k_z$-dependence of the intensity of the scattered X-rays (defined as $|\,E\,|^2$, where $E$ is the amplitude of the scattered radiation) around the (0,0,1) point in reciprocal space. Show that there is a peak at (0,0,1) and that the peak width is finite. How is the width related to the size of the crystal ($N_z$)? How does the height of the peak depend on $N_z$? How does the integrated intensity behave as a function of $N_z$? Assume $N_z \gg 1$.

(b) What happens to the line width and the peak height if we have a collection of small crystallites, which are all oriented parallel to each other, but their positions are random?

## 1.15 Problem: Random Displacement

Consider an "imperfect" crystal, where the $n$th atom is displaced from its ideal position, $R_n$, by a random vector $S_n$. We will look at the diffraction

---

[6] See Problem 1.6.

peak at around the reciprocal lattice vector $G$. Assume that the displacement at site $n$ is $| \, S_n \, | \ll a$, where $a$ is the average lattice spacing. Show that the diffraction peak intensity will change little for small $G$, but will disappear at larger $G$! What is the shape of the envelope (peak height versus $G = | \, G \, |$)?

## 1.16 Problem: Vacancies

Vacancies are missing atoms in an otherwise near-perfect crystal. Since they create disorder and increase the entropy, vacancies are always present at nonzero temperatures. How will the X-ray diffraction of a crystal change due to a small amount of vacancies?

## 1.17 Problem: Integrated Scattering Intensity

Assume that in a scattering experiment the amplitude of scattered radiation is

$$A = A_0 \sum_n e^{iKR_n} \; , \tag{I.1.6}$$

where $K$ is the scattering wavevector and $R_n$ is the position of atom $n$. The intensity is given by $I = | \, A \, |^2$. Show that the integrated intensity,

$$I_{\text{tot}} = \frac{\int d^3 K \, I}{\int d^3 K} \; , \tag{I.1.7}$$

is independent of the atomic positions!

# 2 Interatomic Forces, Lattice Vibrations

Solids can be classified according to the dominant contribution to their cohesive energy: van der Waals, ionic, hydrogen-bonded, covalent solids, and metals. In principle, the total energy of a solid depends on the coordinates and velocities of the atomic nuclei and on the coordinates and velocities (or, in quantum mechanics, the wavefunctions) of the electrons. However, in the *adiabatic approximation* the potential energy is expressed as $U = U(r_1, r_2, \ldots r_i, \ldots)$, where the $r_i$ are the atomic positions, with no explicit dependence on their electronic states. When the nuclei are at equilibrium positions $(R_i)$, the energy $U$ equals the *binding energy* or *cohesive energy* of the solid.

For van der Waals solids the potential energy can be quite well calculated by summing up the pair potential of interacting atoms. For two atoms the interaction is described by the *the Lennard–Jones potential*:

$$\Phi_{ij} = \frac{A}{r_{ij}^6} + \frac{B}{r_{ij}^{12}} , \qquad (\mathrm{I.2.1})$$

or

$$\Phi_{ij} = 4\epsilon \left[ \left( \frac{\sigma}{r_{ij}} \right)^6 + \left( \frac{\sigma}{r_{ij}} \right)^{12} \right] . \qquad (\mathrm{I.2.2})$$

The energy of the lattice can be expressed with the lattice sums $A_6 = \sum_j p_{ij}^{-6}$ and $A_{12} = \sum_j p_{ij}^{-12}$ such that

$$U = 2N\epsilon \left[ A_6 \left( \frac{\sigma}{d} \right)^6 + A_{12} \left( \frac{\sigma}{d} \right)^{12} \right] . \qquad (\mathrm{I.2.3})$$

For ionic solids the Coulomb potential, supplemented with a strongly repulsive core force, is used:

$$\Phi_{ij} = \frac{\pm e^2}{r_{ij}} + \frac{B}{r_{ij}^n} , \qquad (\mathrm{I.2.4})$$

where $e$ is the electron charge, and $B$ and $n$ are parameters to describe the repulsive atomic core. The *Madelung energy* is obtained by adding up the Coulomb potentials in the crystal

$$U = \frac{e^2}{d}\frac{1}{2}\sum_{i,j\ i\neq j}\frac{\pm 1}{p_{ij}} = N\frac{e^2}{2d}\sum_{j}\frac{\pm 1}{p_{0j}} = -MN\frac{e^2}{d}\ ,\qquad (I.2.5)$$

where $N$ is the number of primitive unit cells in the crystal, $d$ is the equilibrium nearest-neighbor distance between the centers of the positive and negative ions, and $p_{ij}$ is the distance between the $i$th and $j$th ion, measured in units of $d$. The $\frac{1}{2}$ factor compensates for double counting the ion pairs as the sum is performed, and the $\pm 1$ represents the attractive and repulsive contributions. The *Madelung constant* is the coefficient $M$ in Eq. I.2.5.

In covalent and hydrogen-bonded materials, the calculation of the cohesive energy is much more complicated. In general, the potential energy cannot be calculated as a sum over pair potentials acting between atoms. For simple metals, an estimate of the cohesive energy can be obtained by balancing the kinetic energy of the electrons against the Coulomb attraction between the atomic cores and the conduction electrons. To calculate the Coulomb energy, the electrons are treated as a uniform negative background; in first approximation the kinetic energy is obtained by solving a simple "particle in a box" quantum mechanics problem. This procedure yields

$$U = N\left(\frac{\alpha}{r^2} - \frac{\beta}{r}\right)\ ,\qquad (I.2.6)$$

where $r$ is the average distance between the electrons.[1]

*Lattice vibrations* are elastic waves propagating within crystals. *Phonons* are quantized elastic waves. The expression $U = U(r_1, r_2, \ldots r_i, \ldots)$ can be expanded around the equilibrium position of the atoms by $r_i = R_i + u_i$, where $R_i$ represents the equilibrium position and $u_i$ is the displacement vector. We obtain the harmonic expansion:

$$U(u_1, u_2, \ldots) = U_0(R_1, R_2, \ldots) + \frac{1}{2}\sum_{ij} u_i D_{ij} u_j\ ,\qquad (I.2.7)$$

where the tensor $D_{ij}$ is obtained from the second derivatives of the potential. Quite often it is enough to keep only the nearest-neighbor terms in the above summation.

The equation of motion is

$$M_i\frac{d^2 u_i}{dt^2} = \sum_{j} D_{ij} u_j\ ,\qquad (I.2.8)$$

where $M_i$ is the mass of the $i$th atom. The solution is searched in the form of a lattice wave,

$$u_i = \epsilon_s e^{ikR_i} e^{i\omega t}\ ,\qquad (I.2.9)$$

where the three $\epsilon_s(k)$ ($s = 1, 2, 3$) are the *polarization vectors* of the vibration. With one atom per unit cell (every $M_i = M$), in the harmonic approximation,

---

[1] See Ashcroft and Mermin [1] p. 410.

the solution to the equation of motion is reduced to solving a simple linear set of equations:

$$D(k)\epsilon(k) = \lambda(k)\epsilon(k) , \tag{I.2.10}$$

where $D(k) = \sum_j D_{\ell j} e^{ikR_j}$ (independent of $\ell$) is the *dynamical matrix*.

In one dimension, for nearest-neighbor-only interactions, Eq. I.2.7 simplifies to

$$U(u_i) = U_0(R_i) + k \sum_i (u_i^2 - u_i u_{i+1}) , \tag{I.2.11}$$

and the equation of motion is also much simpler.

The general solution of the equation of motion provides the *phonon dispersion* or *phonon spectrum* $\omega = \omega(k) = \sqrt{\lambda(k)/M}$. A continuous set of $\omega(k)$ values is a *phonon branch*. The long wavelength (or small wavenumber, $|k| \ll \pi/a$, where $a$ is a lattice spacing) vibrations are sound waves. The phonon branches that start from $\omega = 0$ at $k = 0$ are the *acoustic phonons*. It is easy to show that for acoustic phonons at small $k = |k|$ the phonon frequencies are proportional to $k$ (although the constant of proportionality $c$, the *sound velocity*, may depend on the direction of propagation).

If the wave vector $k$ is along appropriate symmetry axes of the crystal, then one of the polarization vectors will point parallel to $k$ (corresponding to the *longitudinal mode*), while the other two are perpendicular (*transverse modes*). For general directions of $k$, the concept of longitudinal and transverse modes is only approximately valid. For $p > 1$ atoms in the primitive unit cell, the phonon spectrum will have more branches, including the $p - 1$ higher-frequency *optical phonons*. The total number of possible values of $k$ are fixed by the periodic boundary condition as $N_k = N$, where $N$ is the number of primitive unit cells in the crystal (for simple crystals with one atom per unit cell, $N$ is the number of atoms). In three dimensions, for a system of $N^*$ atoms the total number of possible phonon modes is always $3N^* = 3pN$.

At high-symmetry points in the Brillouin zone the calculation of the phonon mode frequencies is much simpler than finding the general solution. For example, in one dimension a zone boundary ($k = \pi/a$) mode corresponds to neighboring atoms oscillating with opposite phases. With this in mind, the equation of motion for the N atom can be reduced to that of a two-atom problem.

Since atoms are massive, a fairly accurate picture of phonons can be obtained without using quantum mechanics. A somewhat oversimplified transition to the "quantum world" is provided by the correspondence principle: The phonon energy is $E = \hbar\omega$ and the momentum is $p = \hbar k$. Note that, loosely speaking, the amplitude of the classical vibrations with wavenumber $q$ and frequency $\omega$ corresponds to the number of phonons in the $q$, $\omega$ state, $n_{q\omega}$. More accurately, the expectation value of the amplitude, $\langle A_{q,\omega} \rangle$, and the expectation value of the number of phonons $\langle n_{q\omega} \rangle$ are related by

$$\langle n_{q\omega} \rangle = \frac{2M\omega}{\hbar} \langle A_{q\omega} \rangle^2 , \tag{I.2.12}$$

where $M$ is the mass of the atoms. Remember: The classical energy of the oscillation, $\frac{1}{2}M\omega^2 A^2$, has nothing to do with the phonon energy!

Two convenient models are frequently used in connection to lattice vibrations. The *Einstein model* is appropriate for the optical modes; it consists of independent oscillators with the same resonance frequency, set equal to the frequency of a typical optical phonon. The total number of oscillators equals $N$, the number of optical phonon modes around that frequency. The *Debye approximation* replaces the density of states for each of the three acoustic modes with the density of states corresponding to the low-$k$ part of the spectrum, where the phonon frequency is proportional to the wavenumber, $\omega = ck$, with $c$ being the sound velocity. Furthermore, the Brillouin zone is replaced with a sphere of radius $k_D$, the *Debye wavenumber*, so that the total number of states within this sphere is equal to the total number of states in the Brillouin zone, $N$. The upper cutoff frequency is the *Debye frequency*, $\omega_D = ck_D$; the corresponding temperature is the *Debye temperature* $\Theta_D = \hbar\omega_D/k_B$.

As we will discuss later, at temperatures much higher than the Debye temperature the number of phonons increases as $N_{ph} \sim T/\Theta_D$. The *Lindemann melting formula* uses the temperature dependence of the phonon number and Eq. I.2.12 to estimate the melting point of solids:

$$T_m = \frac{(A_m)^2}{9\hbar^2} M k_B \Theta_D^2 \,, \tag{I.2.13}$$

$A_m = 0.2 - 0.4\, r_s$ is the amplitude of the thermally excited oscillation at the melting point, $r_s$ is the average size of the unit cell, and $M$ is the mass of the atoms.

*Neutron scattering* is the best way to explore the phonon spectrum, but limited information can also be obtained from *optical spectroscopy* (see introductory text in Chapter 8) and other spectroscopic methods. To tackle the task of presenting a multivalued function of three independent variables (the three components of $k$), it is common to show the phonon dispersion only along a few symmetry directions of the crystal.

In metals the lattice deformations influence the motion of the electrons and vice versa. The *electron–phonon interaction* leads to a damping of the lattice vibration, or finite *phonon lifetime*. The finite phonon lifetime appears as a finite energy width in the scattering experiment. For the electrons, an important contribution to the finite electrical resistance (see introductory text in Chapter 7) is due to phonons; however, the electron–phonon interaction can lead to a total destruction of the metallic conductivity, like in the case of superconductivity or charge density waves.

In the harmonic approximation the compressibility is independent of the temperature, and the the linear thermal expansion coefficient is zero.[2]    To

---

[2] The compressibility of solids is defined as the ratio between the volume change in response to the change in pressure: $\kappa = -\frac{1}{V}dV/dp$. The thermal expansion

understand the thermal expansion, one has to go beyond the harmonic expansion. Anharmonicity also leads to the volume dependence of phonon frequencies. The *Grüneisen parameter*, $\gamma = -d\ln\omega/d\ln V$, describes this effect for small changes in volume. For insulators, the approximate formula $\alpha = \kappa\gamma c_V/3$ points to the intimate relationship between the two manifestations (i.e., thermal expansion and phonon frequency change) of the anharmonicity.

With the anharmonic terms in the potential, the lattice waves (Eq. I.2.9) are not exact solutions of the equation of motion. However, the lattice waves and the phonon concept can be saved if the anharmonic terms are viewed as a source of interaction between the phonons. In this description, phonons have finite lifetimes, and they can decay into other phonons, as long as their energies and momenta (wavenumbers) are conserved.

## 2.1 Problem: Madelung Constant

Calculate the Madelung constant for the $Rb_3C_{60}$ compound (see Problem 1.5). Assume that the $3e^-$ charges are uniformly distributed over each $C_{60}$ molecule and that the $Rb^+$ ion is pointlike.[3]

## 2.2 Problem: NaCl Bulk Modulus

For NaCl the bulk modulus ($B = -V\partial P/\partial V$, where $P$ is the pressure, $V$ is the volume) is $B = 2.4 \times 10^{11}$ dyn/cm$^2$. The equilibrium (zero pressure) distance between the Na and Cl ions is $d = 2.82$ Å, respectively. Assume that the interaction between the ions is described by the potential

$$\Phi_{ij} = \frac{\pm e^2}{r_{ij}} + \frac{\alpha}{r_{ij}^n} \ . \tag{I.2.14}$$

Using the given values, calculate $\alpha$ and $n$.

## 2.3 Problem: Madelung with Screened Potential

Calculate the Madelung constant $M$ for a linear chain numerically and analytically. With a personal computer, calculate $M$ for the NaCl crystal by using the "screened Coulomb" potential $U = \pm e^2/re^{-\alpha r}$. Discuss what happens in the $\alpha \to 0$ limit.

---

is $\alpha = \frac{1}{3V}dV/dT$, where $T$ is the temperature. These concepts are discussed in detail in Chapter 6.

[3] Note that the Madelung sum converges very slowly, and make sure that your computer procedure works on NaCl, or other structures, where the Madelung constant is known.

## 2.4 Problem: Triple-axis Spectrometer

A triple-axis neutron spectrometer[4] consists of three independently controlled axes of rotation for the sample, monochromator, and analyzer crystals, as shown in Figure I.2.1.

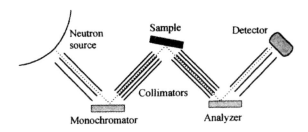

**Fig. I.2.1.** Schematic illustration of a neutron scattering spectrometer. The beam is first diffracted from the monochromator, then it hits the sample. Neutrons scattered from the sample are diffracted by the analyzer crystal and finally reach the detector. The three axes of rotations allow for selection of well defined incident and scattered wavevectors at the sample. Triple-axis spectrometers are built for inelastic scattering, but they may also be used to determine lattice spacings.

Inelastic neutron scattering is performed with a beam neutrons of wavelength $\lambda = 2.502 \pm 0.002$ Å The analyzer crystal provides a wavelength selection with an accuracy of $\delta\lambda/\lambda = 10^{-3}$. What is the the energy resolution of the spectrometer in this configuration?

## 2.5 Problem: Phonons in Silicon

Determine the sound velocity of the transverse acoustic modes in silicon, using the measured phonon spectrum (see, for example, Landolt–Börnstein [21] vol. 17, p. 370). Is this sound velocity isotropic?[5]

## 2.6 Problem: Linear Array of Emitters, Phonons

Consider a linear array of $N$ identical radiation sources, placed at positions $x_n = na$. Each source emits radiation at frequency $\omega$, wavenumber $k$, amplitude $A$ and the sources are coherent with zero phase difference. We detect the radiation with intensity $I_{detector}$ (defined as the time-averaged amplitude

---

[4] See, for example, Ibach and Lüth [4] p. 45.
[5] The phonon spectrum in Ref. [4] p. 57 is incomplete; for the (1 1 0) direction one of the acoustic modes is missing.

square) with a detector positioned along the line of the array, far away from the sources. Assume that $N$ is large.

Introduce a time-dependent modulation to the position of the sources, described by $u_n(t) = u_0 \cos(qan - \Omega t)$. The detector is capable of analyzing the frequency of the incident radiation as well as the intensity. At what frequencies and wavenumbers will strong radiation be detected?

## 2.7 Problem: Long Range Interaction

Particles in a linear chain are located at positions $R_j = aj$, where $a$ is the lattice spacing and $j$ is an integer. The coupling between the particles is represented by springs. The spring constant between the $i$th and $j$th particles depends on the distance between the particles: $D_{i,j} \equiv D(R_i - R_j)$.

What is $D_{i,j}$ if the phonon dispersion is a perfect straight line, $\omega = c\,|\,k\,|$?

## 2.8 Problem: Mass Defect

Calculate the eigenfrequency of a linear chain with a mass defect (isotope) at position $n = 0$, assuming a localized mode, described by[6]

$$s_n = s_0 e^{-k(\omega)|n| - i\omega t} \tag{I.2.15}$$

for the displacement of the $n$th atom, $s_n$. The coupling between the atoms is represented by nearest-neighbor springs of spring constant $\kappa$ (see Figure I.2.2).

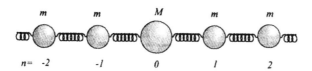

**Fig. I.2.2** Linear chain of coupled atoms, with an isotope at the position $n = 0$

## 2.9 Problem: Debye Frequency

A monoatomic, cubic material has lattice spacing of $a$. The sound velocity for longitudinal and transverse phonons is approximately equal, $c_T = c_L = c$, is isotropic, and the highest phonon frequency is $\omega^*$. What is the Debye frequency?

---
[6] This Ansatz is suggested by Ibach and Lüth [4] p. 61.

## 2.10 Problem: Vibrations of a Square Lattice

A simple, monoatomic, two-dimensional square lattice is modeled by balls and springs. The lattice spacing is $a$, the mass of the balls is $m$, the spring constant of the nearest-neighbor springs (along the edges of the squares) is $k_1$, and the spring constant between the next-nearest neighbors (along the diagonal of the squares) is $k_2$. All other interactions are negligible.

What is the frequency of the $k = (\frac{\pi}{a}, 0)$ mode oscillations? (*Warning:* The complete solution of the equations of motion is time-consuming and *not* necessary for the answer.)

## 2.11 Problem: Grüneisen Parameter

Calculate the Grüneisen parameter, $\gamma = -\partial \ln \omega / \partial \ln L$, for a linear chain of length $L$, lattice spacing $a$, and only nearest-neighbor interactions. Assume that the interaction potential has the form of $U(x) = U_0 + \frac{1}{2}\kappa x^2 + \lambda x^3$, where $x = d - a$, and $d$ is the distance between nearest-neighbors.

## 2.12 Problem: Diatomic Chain

For a diatomic linear chain, the phonon dispersion relation $\omega(q)$ has two branches, according to the $+$ and $-$ sign in the equation below:

$$\omega^2 = f\left(\frac{1}{M_1} + \frac{1}{M_2}\right) \pm f\left[\left(\frac{1}{M_1} + \frac{1}{M_2}\right)^2 - \frac{4}{M_1 M_2}\sin^2\frac{qa}{2}\right]^{1/2} . \quad \text{(I.2.16)}$$

There are two atoms in the unit cell with masses $M_1$ and $M_2$, and the force constant of the nearest-neighbor interaction is given by $f$. In Figure I.2.3 the frequency is plotted for various values of $\alpha = M_1/M_2$, with the effective mass, $\mu = M_1 M_2/(M_1 + M_2)$, kept constant.

(a) Calculate the sound velocity.

(b) Show that for $M_1 = M_2$, the result is equivalent to the dispersion curve of the single-atom chain.

## 2.13 Problem: Damped Oscillation

In a linear chain of lattice spacing $a$, particles of mass $m$ are connected by first-neighbor springs of spring constant $\kappa$. In addition to the elastic forces, each particle is subjected to a damping force $F = -\Gamma \dot{s}_n$, where $s_n$ is the displacement of the $n$th particle from the equilibrium position, and $\dot{s}_n$ is its velocity. How does the damping change the frequencies $\omega = \omega(k)$, and what is the relaxation time of the modes? Assume $\Gamma^2 \ll \kappa/m$ and discuss the $k \approx \pi/a$ and $k \approx 0$ modes separately.

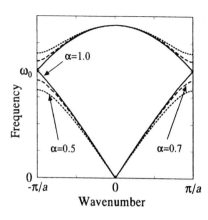

**Fig. I.2.3.** Phonon dispersion relation for a diatomic chain for $\alpha = 1.0$ (solid line), $\alpha = 0.7$ and $\alpha = 0.5$. The frequency scale is set by $\omega_0^2 = f/\mu$.

## 2.14 Problem: Two-Dimensional Debye

A phonon mode in a two-dimensional hexagonal lattice of lattice spacing $a = 3$ Å has isotropic sound velocity of $c = 10^3$ m/sec. What is the Debye frequency, $\omega_D$? (Provide a numerical value in $\sec^{-1}$.)

## 2.15 Problem: Soft Optical Phonons

A linear chain consists of polarizable molecules which are separated by lattice spacing $a$. The molecules are fixed to their position, but they have an internal degree of freedom described by the equation of motion

$$\frac{\partial^2 p}{\partial t^2} = -\omega_0^2 p + E\alpha\omega_0^2 , \qquad (I.2.17)$$

where $p$ is the electric dipole moment of the molecule (assumed to be parallel to the chain), $E$ is the local electric field, and $\alpha$ is the polarizability. Each molecule feels the electric field of the others. The system is at zero temperature, and quantum effects are small. Calculate and plot the dispersion curve $\omega(k)$ for small amplitude polarization waves (optical phonons). Discuss the behavior of $\omega(k = 0)$ as a function of $\alpha$?

## 2.16 Problem: Soft Phonons Again

Same as Problem 2.15, but the dipole moment vectors are perpendicular to the chain. Assume that all dipole moment vectors are in the same plane; disregard the possibility of the "helical" arrangements.

# 3 Electronic Band Structure

Calculating the electronic band structure requires solving the Schrödinger equation in the presence of a fixed and perfectly periodic potential due to the ions, while taking into account the Coulomb repulsion between the electrons. The first part of the job is itself difficult enough, although with sufficient computing power a numerical solution of high accuracy can always be produced. On the other hand, a general treatment of the electron–electron interactions is impossible. The art of band structure calculations is in finding the best approximate methods.

This collection of problems does not provide an opportunity to practice this art. Most of the problems here deal with a single electron (or a set of non-interacting electrons) in a periodic potential. Two extreme cases, the *nearly free-electrons* and the tightly bound electrons (or *tight-binding approximation*), are discussed in detail.

The cornerstone of band structure calculations is the *Bloch condition*: The wavefunctions have to satisfy

$$\psi_{\boldsymbol{k}}(\boldsymbol{r} + \boldsymbol{R}) = e^{i\boldsymbol{k}\cdot\boldsymbol{R}}\psi_{\boldsymbol{k}} \, , \tag{I.3.1}$$

where $\boldsymbol{R}$ is an arbitrary lattice vector. The possible values of the wavenumbers ($\boldsymbol{k}$) are set by the periodic boundary condition, so that within the first Brillouin zone the total number of different $\boldsymbol{k}$ vectors is equal to the number of primitive unit cells (in simple crystals, the number of atoms) in the solid. *Bloch functions*

$$\psi_{\boldsymbol{k}}(\boldsymbol{r}) = u_{\boldsymbol{k}}(\boldsymbol{r})e^{i\boldsymbol{k}\cdot\boldsymbol{r}} \, , \tag{I.3.2}$$

or *Wannier functions*

$$\psi_{\boldsymbol{k}}(\boldsymbol{r}) = \sum_i \phi(\boldsymbol{r} - \boldsymbol{R}_i)e^{i\boldsymbol{k}\cdot\boldsymbol{R}_i} \, , \tag{I.3.3}$$

are two possible ways to satisfy the Bloch condition. In the Bloch functions, $u_{\boldsymbol{k}}(\boldsymbol{r})$ is a periodic function such that $u_{\boldsymbol{k}}(\boldsymbol{r}+\boldsymbol{R}) = u_{\boldsymbol{k}}(\boldsymbol{r})$ for any lattice vector $\boldsymbol{R}$. There are no particular restrictions on $\phi(\boldsymbol{r})$ in the Wannier functions.

In the nearly free-electron approximation the lattice potential is treated as a perturbation. The starting point of the calculation is a set of electronic wavefunctions in the form of plane waves

$$\psi_k^0 = e^{ik \cdot r} \, , \tag{I.3.4}$$

and the energies are $E_k = \frac{\hbar^2}{2m} k^2$.

In the presence of an external potential $V(r)$ we search for the wavefunction in the form of *linear combinations of plane waves*:

$$\psi_k = \sum_G C_{k-G} e^{i(k-G)r} \, , \tag{I.3.5}$$

where the $G$'s are reciprocal lattice vectors. Inserting this wavefunction into the Schrödinger equation leads to a set of linear equations,

$$[E - E(k-G)] C_{k-G} = \sum_{G'} C_{k-G'} V_{G-G'} \, , \tag{I.3.6}$$

where $E(k-G) = (k-G)^2/2m$, and $V_G$ is the Fourier component of the lattice potential $V(r)$. This way the problem is reduced to solving a set of algebraic equations, as discussed in detail by Ziman [3] p. 80, Ashcroft and Mermin [1] p. 154, Kittel [2] p. 167, or Ibach and Lüth [4] p. 111, and others. Among other things, the results of this calculation tell us the *energy gap* at the Brillouin zone boundary. For example, if we assume that only the $G - G' = G_0$ Fourier component of $V_{G-G'}$ is nonzero, Eq. I.3.6 is easy to solve, and the result is

$$E^{\pm}(k) = \frac{1}{2}(E_k + E_{k-G_0}) \pm \frac{1}{2}\sqrt{(E_k - E_{k-G_0})^2 + 4V_{G_0}^2} \, , \tag{I.3.7}$$

where $E_k = \frac{\hbar^2}{2m} k^2$.

The tight-binding approximation starts in the other extreme: The wavefunctions are localized to the atoms; the solution of the Schrödinger equation for the single-atom problem, $\Phi_n(r)$, is used in the Wannier functions (Eq. I.3.3)

$$\psi_{n,k}(r) = \sum_i \Phi_n(r - R_i) e^{ik \cdot R_i} \, . \tag{I.3.8}$$

For each atomic energy level $n$ (or more generally $n, l, m$, where $l$ and $m$ are the angular momentum values) we can substitute Eq. I.3.8 into the Hamiltonian and obtain the energy spectrum.[1] For a one-dimensional lattice of lattice spacing $a$ the result is

$$E(k) = E_n - \frac{\beta + 2\gamma \cos ka}{1 + 2\alpha \cos ka} \, , \tag{I.3.9}$$

where the nearest-neighbor overlap integrals are, in good approximation:

---

[1] The general solution is discussed, for example, by Ashcroft and Mermin [1] p. 181.

$$\gamma = -\int dr \ \Phi^*(r)\Phi(r-a)\Delta U(r) \ ,$$

$$\beta = -\int dr \ \Phi^*(r)\Phi(r)\Delta U(r) \ , \qquad (\text{I.3.10})$$

$$\alpha = \int dr \ \Phi^*(r)\Phi(r-a) \ .$$

Here $\Delta U(r)$ represents all atomic potentials, except the one located at position $r = 0$. Since the $\alpha$ integral is small, the general form of the band structure can be further simplified to

$$E(k) = E_0 - 2(\gamma)\cos ka \ . \qquad (\text{I.3.11})$$

In general, the *energy band* is the collection of a continuous set of $E(k)$ values, obtained by solving the Schrödinger equation. The tight-binding approximation illustrates the relationship between the energy bands and the original atomic energy levels very well.

There are a few simple rules to help us to "guess" the band structure.

- Constant energy contour plots are perpendicular to the Brillouin zone boundary.
- Along high symmetry lines of the Brillouin zone, $E(k)$ is perpendicular to the zone boundary.
- The energy curve is symmetric under the inversion $k \to -k$ .

Exceptions to the first two principles occur is two or more bands are degenerate at the Brillouin zone boundary. The last rule follows from the time reversal symmetry of the Hamiltonian, and it is not necessarily valid in the presence of a magnetic field.

After solving the one-electron problem, we have to deal with the other $6\times 10^{21}$ electrons (a typical number of electrons in 1 g of a solid). The simplest way to handle this immense problem is to invoke the Pauli principle, which forbids more than two electrons from occupying a single $k$ state. At zero temperature this will lead to the familiar *rigid band* picture: noninteracting electrons in a set of different $k$ and spin states. We refer the reader to other textbooks for discussions of the *Fermi liquid theory* which lends some justification to this picture.

Even an approximate band structure calculation can give us an idea about an important property of the material: Do the electrons fill some bands completely, leaving others empty, or are there partially filled bands? The answer to this question determines if the material is a *metal* or an *insulator*. At this level the metallic character has nothing to do with the "extension" of the atomic wavefunction; partially filled tight-binding bands indicate a metallic behavior.[2] The *Fermi surface* is the constant energy surface separating empty

---

[2] In a more sophisticated approach, when the interaction between the electrons, phonons, and impurities are taken into account, one often finds that electrons in

and filled electronic states in the Brillouin zone, and the *Fermi energy* is the corresponding energy. Strictly speaking, the Fermi surface and Fermi energy are defined at zero temperature (at finite temperature there is no sharp separation between the filled and empty states). For most common metals the thermal excitation energies are small compared to the Fermi energy (in other words, the *Fermi temperature*, $T_F = E_F/k_B$, is much larger than room temperature). At a typical temperature $T$ the electrons in a narrow energy range $k_B T$ around the Fermi energy determine many of the properties of metals.

If the electron experiences the lattice potential and an external electric and/or magnetic field simultaneously, then the Schrödinger equation cannot be solved exactly. However, for weak fields the *quasiclassical approximation* is valid.[3] The velocity of a wave packet consisting of electron waves can be regarded as the electron velocity:

$$v = \frac{\partial \omega}{\partial k} = \frac{1}{\hbar} \frac{\partial E}{\partial k} . \qquad (I.3.12)$$

Furthermore, under the influence of electromagnetic forces the wavenumber becomes time-dependent, similar to the momentum in classical mechanics:

$$\frac{\partial k}{\partial t} = -\frac{e}{\hbar} \left( E + \frac{1}{c} v \times H \right) \qquad (I.3.13)$$

where $E$ is the electric field and $H$ is the magnetic field. These equations are used extensively to describe the transport properties of the electronic systems as will be discussed in Chapter 7.

Free-electrons in the presence of a homogeneous magnetic field $H$ are forced into circular orbits resulting in the quantization of energy levels described as $E_\nu = \hbar \omega_c = \hbar e H / mc$ (*Landau levels*). Similar quantization occurs for electrons in a crystal, and the quasiclassical equations I.3.12 and I.3.13 can be used to to derive the relationship (valid for large $\nu$ only):

$$E_{\nu+1} - E_\nu = \frac{2\pi e H}{\hbar c} \left( \frac{\partial A(E, k_z)}{\partial E} \right)^{-1} , \qquad (I.3.14)$$

where the electron momentum parallel to the magnetic field is denoted by $k_z$, and $A$ is the area of the electron orbit cut from a constant energy surface in the $k$ space in the direction perpendicular to the field. When the magnetic field is varied in a metal, the Landau levels repeatedly cross the Fermi energy, and this leads to oscillations in the physical properties. If plotted as the function of the inverse field, the oscillation is periodic with the period of

---

narrow bands do not survive in their extended state, and the material becomes an insulator. This transition is, however, outside of the scope of the present chapter; related issues will be discussed in Chapter 9.

[3] The validity range of the quasiclassical approximation is discussed by many textbooks. See, for example, Ashcroft and Mermin [1] pp. 214–220.

$$\Delta\left(\frac{1}{H}\right) = \frac{2\pi e}{\hbar c}\frac{1}{A(E,k_z^{ext})}\,, \tag{I.3.15}$$

where $A(E,k_z^{ext})$ is the extremal area of the Fermi surface, as defined in several textbooks. In two-dimensional electronic systems the energy does not depend on $k_z$ and the quantization of the energy levels leads to very pronounced variation of the transport properties (*integer quantum Hall effect*).

In solids, due to the periodic potential acting on the electron, the energy does not depend on the wavenumber in a quadratic fashion as it did for free-electrons. However, there will be points in the Brillouin zone where the $E(k)$ function is locally quadratic in $k - k_0$ (naturally, these are the energy minima and maxima; in all likelihood, one of these is at $k_0 = 0$). For electrons moving in the neighborhood of one of these points the concept of *effective mass* is helpful: many of the free-electron properties can be recovered, except the mass is replaced by the effective mass tensor. For energy minima the effective mass tensor is defined as

$$\mathbf{M} = \hbar^2(\mathbf{Q})^{-1}\,, \tag{I.3.16}$$

where the tensor $\mathbf{Q}$ is the second derivative of the energy dispersion:

$$[\mathbf{Q}]_{ij} = \frac{\partial^2 E(k)}{\partial k_i\,\partial k_j}\,. \tag{I.3.17}$$

By selecting the direction of the coordinate axes in a proper way, $\mathbf{Q}$ – and consequently $\mathbf{M}$ – can always be made diagonal. In the most general case $E(k)$ becomes

$$E(k) = \frac{\hbar^2 k_x^2}{2m_1} + \frac{\hbar^2 k_y^2}{2m_2} + \frac{\hbar^2 k_z^2}{2m_3}\,. \tag{I.3.18}$$

When the concept of effective mass works, the electronic density of states, the specific heat, and the motion of the electron under the influence of electric and magnetic fields can be treated in a fashion similar to free-electrons. For example, the effective mass entering to the specific heat expression (see Chapter 6) is

$$m' = [\det | M_{ij} |]^{1/3}\,, \tag{I.3.19}$$

whereas the cyclotron effective mass is defined as

$$m^* = \left[\det\frac{| M_{ij} |}{M_{zz}}\right]^{1/2} \tag{I.3.20}$$

(assuming that the magnetic field points to the $z$ direction).

In the proximity of an energy maximum of $E(k)$, the second derivative $\mathbf{Q}$ is negative. However, we have to keep in mind that states near the maximum of $E(k)$ – that is, near the top of an energy band – come into play if the band is nearly full; consequently it is more practical to talk about the few missing electrons than to keep track of all others. A missing electron with negative mass and negative charge leaves behind a positive mass and positively charged *hole* state. The hole effective mass tensor is defined as $\mathsf{M} = -\hbar^2(\mathsf{Q})^{-1}$, so that the components $m_1$, $m_2$, and $m_3$ are still positive.

The effective mass is extensively used to describe the behavior of *semiconductors*, where a relatively small energy gap separates the filled and empty electronic states. Naturally, the thermally excited electrons will move to the minimum of the upper band, and the empty states are going to be near the top of the lower band. In metals, however, the usefulness of the concept is limited. Although the calculation of the effective mass tensor at an arbitrary *k* point in the Brillouin zone is perfectly possible, neither the thermodynamic nor the transport properties can be conveniently treated in this manner.

## 3.1 Problem: Nearly Free Electrons in One Dimension

Electrons of mass $m$ are confined to one dimension. A weak periodic potential, described by the Fourier series $V(x) = V_0 + V_1 \cos 2\pi x/a + V_2 \cos 4\pi x/a + ...$, is applied.

(a) Under what conditions will the nearly free-electron approximation work? Assuming that the condition is satisfied, sketch the three lowest energy bands in the first Brillouin zone. Number the energy bands (starting from one at the lowest band).

(b) Calculate (to first-order) the energy gap at $k = \pi/a$ (between the first and second band) and $k = 0$ (between the second and third band).

## 3.2 Problem: Nearly Free Electrons in Dirac-Delta Potentials

Atoms are arranged in a one-dimensional chain with lattice spacing $a$. Each atom is represented by the potential $V(x) = aV_0\delta(x)$. Determine the energy gaps between the bands, assuming that the nearly free-electron approximation applies.

## 3.3 Problem: Tight-Binding in Dirac-Delta Potentials

Each atom in a one-dimensional chain of lattice spacing $a$ is represented by a potential of $V(x) = aV_0\delta(x)$.

(a) Determine the ground state wavefunction of the single-atom problem.

(b) Calculate the bandwidth of the lowest energy band in the tight-binding approximation.

## 3.4 Problem: Dirac-Delta Potentials

Atoms are arranged in a one-dimensional chain with a lattice spacing of $a$. Each atom is represented by the potential $V(x) = aV_0\delta(x)$.

(a) Show that the electron energy $E$ and wavenumber $k$ satisfy the relationship

$$\cos ka = \frac{\kappa}{K}\sin Ka + \cos Ka \qquad (I.3.21)$$

where $K^2 = 2mE/\hbar^2$ and $\kappa = aV_0$. Determine the coefficient $\alpha$.

(b) Calculate the energy gaps between the bands for weak potential, $V_0 \ll \hbar^2/ma^2$ and compare the result to the gaps obtained in the nearly free-electron approximation.[4]

(c) Calculate the bandwidth of the lowest-energy band for strong potential, $V_0 \gg \hbar^2/ma^2$ and compare the result to the bandwidth obtained in the tight-binding approximation.[5]

## 3.5 Problem: Band Overlap

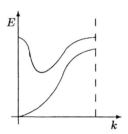

**Fig. I.3.1.** Impossibility of band overlap in a one-dimensional system.

Show that a band overlap (see Figure I.3.1) is not allowed in one dimension.

## 3.6 Problem: Nearly Free Electrons in Two Dimensions

Assume we have a two-dimensional nearly free-electron system on a square lattice of lattice spacing $a$. The Fourier transform of the weak lattice potential is $V(\mathbf{G})$.

We want to investigate the band structure around the $(\pi/a, \pi/a)$ point in the reciprocal lattice. The unperturbed spectrum has a fourfold degeneracy at this point. Only the $\mathbf{G} = (0, 2\pi/a)$ and the $\mathbf{G} = (2\pi/a, 2\pi/a)$ components of $V_{\mathbf{G}}$ are important. Find the gap if $V_{(0,2\pi/a)} = V_0$ and $V_{(2\pi/a,2\pi/a)} = 0$. Also, find the gap if $V_{(0,2\pi/a)} = 0$ and $V_{(2\pi/a,2\pi/a)} = V_1$.

---

[4] see Problem 3.2.
[5] see Problem 3.3.

## 3.7 Problem: Nearly Free Electron Bands

An electron of mass $m$ moves in a square lattice of lattice spacing $a$. Sketch $E(k)$ along the $\Gamma - W$ and $W - X$ lines (see Figure I.3.2) in the limit of a very weak lattice potential.

## 3.8 Problem: Instability at the Fermi Wavenumber

A weak periodic potential in the form of $V(x) = V_0 \cos(2k_F x)$ is created for a one-dimensional electron system ($k_F$ is the Fermi wavenumber). Calculate $E(k)$ for the lowest energy band, and determine the total energy of the system at zero temperature as a function of $V_0$.

## 3.9 Problem: Electrons in 2D Nearly Free Electron Band

An electron of mass $m$ moves in a square lattice of lattice spacing $a$. The nearly free-electron approximation applies.

(a) With one electron per site in the crystal, draw the Fermi surface on the $k_x, k_y$ plane. Is this a metal or an insulator?

(b) With two electrons per site, draw the Fermi surface. Is this a metal or an insulator?

## 3.10 Problem: Square Lattice

Consider two-dimensional electrons subjected to a weak periodic potential coming from a square lattice of spacing $a = 5$ Å. For $k$ vectors far away from the Brillouin zone boundary, the wavefunction can be well described by plane waves. Assume we want to write the wavefunction in the Bloch form, $\psi(k) = e^{ik \cdot r} u(r)$. We consider a state of energy $E$ and wavevector

$$k = \begin{pmatrix} 0.5 \text{ Å}^{-1} \\ 0 \end{pmatrix}. \tag{I.3.22}$$

What will the three lowest energies be at this wavenumber? What are the corresponding $u(r)$ functions? (Note: $\hbar^2/2m = 3.806$ eVÅ$^2$.)

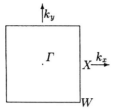

**Fig. I.3.2.** First Brillouin zone.

## 3.11 Problem: Tight-Binding Band in Two Dimensions

Consider electrons on a two-dimensional square lattice in tight-binding approximation, $E = E_0(2 - \cos ak_x - \cos ak_y)$.

(a) Plot a few constant energy lines on the $k_x, k_y$ plane.

(b) Show that most of the constant energy lines cross the Brillouin zone boundary at a right angle.

(c) Sketch $E(\mathbf{k})$ along the $W - X$ line (shown in Figure I.3.2).

## 3.12 Problem: Electrons in 2D Tight-Binding Band

Consider electrons on a two-dimensional square lattice in the tight-binding approximation, $E = E_0(2 - \cos ak_x - \cos ak_y)$.

(a) With one electron per site in this crystal, draw the Fermi surface. Is this a metal or an insulator?

(b) With two electrons per site, draw the Fermi surface. Is this a metal or an insulator?

## 3.13 Problem: Dirac-Delta Potentials in Two Dimensions

An electron of mass $m$ moves in a square lattice of lattice spacing $a$. The lattice potential is represented by Dirac-delta functions; $V(\mathbf{r}) = - \sum V_0 a^2 \delta(\mathbf{r} - \mathbf{R}_n)$.

(a) For the nearly free-electron approximation to work, in what sense must the potential be "weak"? (What is the condition on $V_0$?)

(b) How big is the energy gap between the second and third lowest energy bands?

## 3.14 Problem: Effective Mass

Calculate the effective mass tensor $(M_{ij})$ for electrons in a simple cubic tight-binding band at the center $(\mathbf{k} = \frac{\pi}{a}(0,0,0))$, at the face center $(\mathbf{k} = \frac{\pi}{a}(0,0,1))$, and at the corner $(\mathbf{k} = \frac{\pi}{a}(1,1,1))$ of the Brillouin zone. Discuss the usefulness of the effective mass concept at $\mathbf{k} = \frac{\pi}{a}(0,0,1)$.

## 3.15 Problem: Cyclotron Frequency

If the magnetic field is applied in the $z$ direction, the cyclotron effective mass
is defined as

$$m^* = \left[ \frac{\det |M_{ij}|}{M_{zz}} \right]^{1/2} . \qquad (I.3.23)$$

Calculate the cyclotron frequency for electrons at the Fermi surface in a
nearly empty tight-binding band, described by

$$E = -(E_1 \cos k_x a + E_2 \cos k_y b + E_3 \cos k_z c) , \qquad (I.3.24)$$

and show that the result indeed corresponds to the cyclotron frequency of
free electrons of mass $m^*$.

## 3.16 Problem: de Haas–van Alphen

Magnetic field $H$ is applied perpendicular to the layers of conduction in a
quasi-two-dimensional material. The crystal structure of the layers is tetragonal with lattice spacing $a = 3.5$ Å. The electron energy band within the
layers is well described by the tight-binding approximation, and there is no
energy dispersion in the third direction.

The resistance and other material properties exhibit oscillations as the
field varies. The oscillation is periodic in $1/H$ and the period is $6.1 \times 10^{-8}$
$G^{-1}$. What is the conduction electron density (in units of $cm^{-2}$) for this
material? (Be careful: The effective mass approximation does not work in
this problem!)

## 3.17 Problem: Fermi Energy

We have a two-dimensional hexagonal lattice of lattice spacing $a = 3$ Å, and
one electron per unit cell. If the electrons are considered free within the two-
dimensional plane, what is the Fermi energy $E_F$? (Provide a numerical value
in eV.)

# 4 Density of States

As we will see in Chapter 6, the calculation of the thermodynamic properties calls for evaluating sums of the type $\sum_\alpha F(E_\alpha)$, where $\alpha$ runs over all states of the system. According to the Bloch theorem, these states are characterized by wavenumber $\boldsymbol{k}$, and possibly other quantum numbers denoted here by $\sigma$ (like spin state, band index, polarization vector). The sum over wavenumbers is usually turned into an integral over $\boldsymbol{k}$ values, by introducing a factor of $d^3\boldsymbol{k}/(\Delta\boldsymbol{k})^3$, where $(\Delta\boldsymbol{k})^3 = (2\pi/L)^2 \equiv 8\pi^3/V$ is the volume occupied by one state in the $\boldsymbol{k}$ space. When the quantity $F$ depends on the energy only, the integration is best performed by using the energy as a variable. This line of reasoning is summarized as

$$\sum_\alpha F = \sum_{\boldsymbol{k},\sigma} F = \frac{N_0}{V} \sum_\sigma \int_{BZ} \frac{d^3\boldsymbol{k}}{8\pi^3} F = \frac{N_0}{V} \sum_\sigma \int F(E) g_\sigma(E) dE , \quad \text{(I.4.1)}$$

where BZ indicates integration over the first Brillouin zone.

The quantity $g_\sigma(E)$ is the *density of states* (DOS) for systems where a single nondegenerate quantum state belongs to each $\boldsymbol{k}$ state. For systems where degeneracies exist (like spin $1/2$ electrons in a zero magnetic field or phonons along high symmetry lines of a crystal), a portion of the $\sum_\sigma$ can be trivially performed, and an appropriate factor $Z$ can be included in the DOS: $g(E) = Z g_\sigma(E)$.

A more "physical" definition of the DOS starts with $\nu$ quantum states between $E$ and $E + dE$. The density of states is[1] $g(E) = (1/V) d\nu/dE$. Either way, the calculation of the DOS is based on the fact that the states have a uniform density in $\boldsymbol{k}$ space; this follows from the periodic boundary condition, as discussed in many textbooks.

Using the DOS in the thermodynamic calculations allows us to forget about the details of the band structure and converts complicated momentum integrals into simpler energy integrals. For electrons (or, in general, for fermions with fixed particle number) the DOS at the Fermi energy is of crucial importance. To some extent the DOS is also useful when transport properties are to be investigated.

---

[1] Ziman [3] uses the definition of $D(E) = (1/N) d\nu/dE$; Ashcroft and Mermin [1] prefer $g(E) = (1/V) d\nu/dE$; Kittel [2] has $\mathcal{D}(E) = d\nu/dE$ with no normalization at all. Other textbooks use various permutations of these symbols and definitions.

For simplicity, let us assume that in the energy range of interest there is only a single band (or branch) of excitations with dispersion $E = E(\mathbf{k})$. From the perspective of calculating the density of states, it does not matter if we are considering electrons, phonons, or other quasiparticles. We assume that the total number of states in the Brillouin zone is $ZN_0$, where $N_0$ is the number of primitive unit cells in the solid (in simple solids $N_0$ is the number of atoms or molecules) and $Z$ is the energy degeneracy of a single $\mathbf{k}$ state (for electrons with two spin states, $Z = 2$). Starting from the definition, a few formal mathematical operations lead to

$$g = Z \int_{BZ} \frac{d^3 k}{8\pi^3} \delta(E - E(\mathbf{k})) \,, \tag{I.4.2}$$

where $\delta$ is the Dirac-delta function.

Since the total number of $\mathbf{k}$ states in the Brillouin zone is $N_0$, the energy integral of the DOS satisfies

$$\int g(E) dE = \frac{ZN_0}{V} \,. \tag{I.4.3}$$

The DOS is usually zero below a certain energy $E_1$ and above another energy $E_2$; the difference $W = E_2 - E_1$ is called the *bandwidth*. A crude estimate for the DOS for any system is $g = N_0/VW = n/W$, where V is the volume and $n$ is the number density.

A more practical formula is obtained by further manipulation of Eq. I.4.2:

$$g = Z \int \frac{dS}{8\pi^3} \frac{1}{|(\partial E(\mathbf{k})/\partial \mathbf{k})|} \,, \tag{I.4.4}$$

where the integral is over a constant energy surface and in three dimensions: $|\partial E(\mathbf{k})/\partial \mathbf{k}| = \sqrt{(\partial E/\partial k_x)^2 + (\partial E/\partial k_y)^2 + (\partial E/\partial k_z)^2}$. Equation I.4.4 can be rewritten in terms of the group velocity (Eq. I.3.12) of electrons as

$$g = Z \int \frac{dS}{8\pi^3} \frac{1}{\hbar |v(\mathbf{k})|} \,. \tag{I.4.5}$$

A similar expression works for phonons or other quasiparticles. While Eqs. I.4.4 and I.4.5 are often used in "pencil and paper" calculations, computer programs to calculate the DOS are usually based on the original definition: Draw two closely spaced energy surfaces in reciprocal space and calculate the volume in between.

In one dimension the constant energy surface integral in Eq. I.4.4 is reduced to a simple factor of two (accounting for the positive and negative values of $k$). Therefore, in one dimension we obtain

$$g = \frac{1}{2\pi} \frac{2Z}{|dE(k)/dk|} \,. \tag{I.4.6}$$

When the energy dispersion of the particles is approximated by a simple power law, $E \sim k^n$ ($n = 1, 2$), the calculation of the DOS is straightforward.[2] Table I.4.1 summarizes the typical behavior of the DOS for one-, two-, and three-dimensional systems. These results are more than mathematical curiosities: The quadratic DOS of a phonon gas helps to explain the low-temperature specific heat of solids; the divergence of the DOS in one-dimensional electronic systems is responsible for the instabilities in quasi-one-dimensional charge and spin density wave compounds like $TaS_3$ or the organic compound $TMTSF$-$PF_6$; the constant density of states of the two-dimensional electron gas is the starting point of discussing the quantum Hall effect.

| | $d = 1$ | $d = 2$ | $d = 3$ |
|---|---|---|---|
| $E \sim k$ | const. | $\sim E$ | $\sim E^2$ |
| $E \sim k^2$ | $\sim 1/\sqrt{E}$ | const. | $\sim \sqrt{E}$ |

**Table I.4.1.** Energy dependence of the density of states for massive and massless particles at $d = 1$, 2, and 3 dimensions

Equations I.4.4–I.4.6 suggest singular behavior if $|\partial E(k)/\partial k| = 0$ – for example, when $k$ is close to the Brillouin zone boundary. At the corresponding energies, the density of states develops *van Hove singularities*. The singularity is less and less pronounced as the dimensionality is increased (for an illustration, see Problem 4.11).

The properties of a phonon gas are often calculated in terms of the *Debye approximation*, as discussed in Chapter 2. The corresponding DOS (per unit frequency and volume) for each phonon branch is

$$g_s = \frac{\omega^2}{2\pi^2 c_s^3} = 3n_a \omega^2/\omega_{Ds}^3 \qquad \text{for } 0 < \omega < \omega_{Ds} ,$$

$$g_s = 0 \qquad \text{otherwise,} \qquad (I.4.7)$$

where $c_s$ is the sound velocity of phonon branch $s$, and the frequency $\omega$ was used instead of the energy. The number density of the atoms in the solid is denoted by $n_a = N_a/V$, and the Debye frequency of mode $s$ is $\omega_{Ds}$.

## 4.1 Problem: Density of States

Calculate the density of states for massless and massive particles in $d = 1$, 2, and 3 dimensions, and prove the results shown in Table I.4.1.

---

[2] A power law is a good approximation for the low-energy regime of the band structure. "Massive" particles have quadratic dispersions, and "massless" particles have linear dispersions. A few examples: electrons and magnons with nonzero anisotropy energy are massive; phonons and magnons with zero anisotropy energy are massless.

## 4.2 Problem: Two-Dimensional Density of States

Investigate the density of electron states in a two dimensions for energies close to the bottom or the top of a band. Do not use the nearly free-electron approximation, but instead assume that $E(\boldsymbol{k})$ is simply a smooth function.

## 4.3 Problem: Two-Dimensional Tight-Binding

The density of states for a two-dimensional system of electrons with the "tight-binding" band structure $E(\boldsymbol{k}) = -E_0(\cos k_x a + \cos k_y a)/4$ is shown in Figure I.4.1. Investigate the density of states in the neighborhood of $E = 0$.

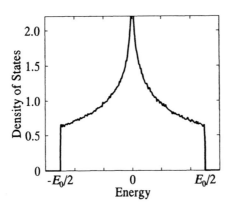

**Fig. I.4.1.** Density of states for the two-dimensional tight-binding band structure. The total area under the curve is normalized to 1.

## 4.4 Problem: Quasi-One-Dimensional Metal

In a quasi-one-dimensional metal, the overlap of the partially occupied electronic orbits is large along one direction, but much smaller in the other two directions. Assume a band that is described by

$$E = -\frac{E_0}{2}\left[A\cos(q_x a) + B\cos(q_y b) + B\cos(q_z b)\right], \qquad (\mathrm{I.4.8})$$

where $q_x$, $q_y$, and $q_z$ are the components of the electron momentum and $A \gg B$. How will the constant energy cuts look on the $q_x, q_y$ plane? Show that for a nearly half-filled band the DOS (and therefore the thermodynamic properties like total energy, total number of electrons, and specific heat) of this system are similar to a truly one-dimensional electron gas! (*Note*: Materials like $TaS_3$ and $NbSe_3$ can be viewed this way.)

## 4.5 Problem: Crossover to Quasi-One-Dimensional Metal

A quasi-one-dimensional electron band is described by

$$E = -\frac{E_0}{2} \left[ A \cos(q_x a) + B \cos(q_y b) + B \cos(q_z b) \right], \qquad (\text{I}.4.9)$$

where $q_x$, $q_y$, and $q_z$ are the components of the electron momentum and $A \gg B$.

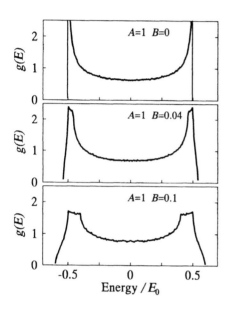

**Fig. I.4.2.** Density of states for quasi−one−dimensional energy bands, as defined by Eq. I.4.9.

Figure I.4.2 shows the DOS at several values of $A$ and $B$. At low band fillings, when the Fermi energy is close to the bottom of the band, the difference between the truly one-dimensional ($B = 0$) and quasi-one-dimensional DOS is significant.

(a) Estimate the value of the electron density when the 1D → 3D crossover happens.

(b) Compare the asymptotic behavior of the 1D band and the 3D band at low energies.

## 4.6 Problem: Phonon Mode of Two-Dimensional System

Let us consider a two-dimensional system with a single-phonon mode of dispersion relation

$$\omega^2 = \omega_0^2(2 - \cos(k_x a) - \cos(k_y a)).$$ 
(I.4.10)

The function given by Eq. I.4.10 is shown in Figure I.4.3.[3]

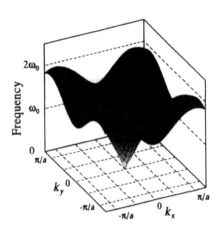

**Fig. I.4.3.** Two-dimensional phonon dispersion given by Eq. I.4.10.

Use a computer to calculate the DOS. Divide the valid range of $k_x$ and $k_y$ values into equal parts. Divide the energy range into equal parts. Count how many $k$ states fall into a given energy interval, and make a histogram.

Using the same scales, also plot the appropriate Debye density of states.

## 4.7 Problem: Saddle Point

In a certain metal the electron energy has a saddle point at $k_0$. In the neighborhood of $k_0$ the energy dispersion can be described by

$$E(k) = E_0 + E^* a^2 k_x^2 - E^* b^2 k_y^2 + E^* c^2 k_z^2 .$$
(I.4.11)

What is the contribution of this region to the DOS? Make a drawing of $g(E)$.

## 4.8 Problem: Density of States in Superconductors

In the BCS theory of superconductivity the excitation spectrum of the single-electron states is

---

[3] This expression is a straightforward generalization of the one-dimensional dispersion formula. It is also the solution of an equation of motion, as discussed by Kittel [2] Problem 4.1, p. 122. The atomic displacements are perpendicular to the plane. Interestingly, the interaction required to obtain this result does *not* correspond to "springs." Rather, it is like the bending of elastic "rods" between the atoms.

$$E(k) = \pm\sqrt{(\epsilon(k) - E_F)^2 + \Delta(k)^2}, \tag{I.4.12}$$

where $\epsilon(k)$ is the electron energy in the absence of superconductivity and $E_F$ is the Fermi energy. With a simplified interaction between the electrons, one obtains $\Delta(k) = \Delta$. Similar expression is obtained for other systems where interactions lead to nonmetallic ground state, like charge density waves, spin density waves, and so on. Calculate the DOS assuming that $E_F$ is not close to a van Hove singularity and $\Delta \ll E_F$.


## 4.9 Problem: Energy Gap

In the presence of a weak periodic potential, $U = U_0 \cos qx$, the energy band of the one-dimensional free-electron gas will develop an energy gap of magnitude $V_0$. The gap will open at wavenumbers $k_0 = \pm q/2$, and around the energy $E_0 = (\hbar^2 q^2)/(8m)$.[4] Calculate the density of states, assuming that the gap is much smaller than $E_0$. (For the purpose of this calculation the two branches of the unperturbed energy spectrum can be approximated by $E \approx E_0 \pm (\hbar/2)v_0(k - k_0)$, where $v_0 = \hbar k_0/m$ is the group velocity of the electrons. The unperturbed density of states is constant, $g_0 = (4/\pi)(k_0/E_0)$.)


## 4.10 Problem: Density of States for Hybridized Bands

If electrons can jump between a broad band [described by $E_c(k)$] and a narrow energy level at $E_0$, then the band structure exhibits signs of *hybridization*: A mixing of electronic states when the $E_c(k)$ and $E_0$ energies are close (see Problem 5.2). In a simple model, the hybridized energy levels can be calculated as

$$E' = (E_c + E_0)/2 \pm \sqrt{(E_c - E_0)^2/4 + \Delta^2} . \tag{I.4.13}$$

Assume that the broad band has $g_c(E_c) = g_0$, a constant DOS over a bandwidth $W = E_2 - E_1$, and $g_c(E_c) = 0$ for $E_c < E_1$ or $E_c > E_2$. For the narrow band the DOS peaks sharply at energy $E_0$, but the total number of states is the same as in the broad band; $g = g_0 W \delta(E - E_0)$. Calculate the DOS for the hybridized band.


## 4.11 Problem: Infinite-Dimensional DOS

Figure I.4.4 shows the DOS for tight-binding electron bands in $d = 1, 2, 3$, and 4 dimensions. The bandwidth is normalized to 1. What is the DOS for $d \to \infty$?

---

[4] The band structure of a one-dimensional electron gas in the presence of weak, periodic potential has been discussed in problems Problems 3.1. and 3.8.

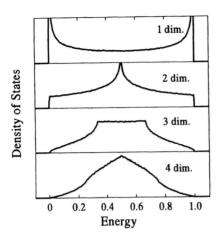

**Fig. I.4.4.** Density of states in the tight-binding approximation for dimensionality $d = 1, 2, 3$, and 4. The energy scale is normalized to the bandwidth, and the DOS is normalized so that the integral under the curves is equal.

## 4.12 Problem: Two-Dimensional Electron Gas

The dimensionality of a system can be reduced by confining the electrons in certain directions. Consider an electron gas in an external potential: $V = 0$ for $|z| < d/2$ and $V = V_0$ for $|z| > d/2$. What is the density of states as a function of energy for $V_0 \to \infty$? (Discuss what happens at low and high energies.)

Assume $d = 100$ Å. Up to what temperatures can we consider the electrons to be two-dimensional? If we can produce a potential of 100 meV and reach a temperature of 20 mK, what is the range of thicknesses feasible for the study of such a two-dimensional electron gas? (*Note*: A two-dimensional electron gas is produced in semiconductor devices and is used for the investigation of the quantum Hall effect as well as other phenomena.)

# 5 Elementary Excitations

Considering the difficulty of handling a large number of interacting particles, many times we have to be happy if we understand only the low-temperature behavior of the solid. The task is to find the *ground state*, only strictly relevant at zero temperature. For example, in sodium metal the ground state of the electrons is a simple Fermi sea filled to the level of one electron per atom; in niobium the electrons make up Cooper pairs and superconduct in the ground state; for iron the ground state of the electrons is ferromagnetic. As these examples illustrate, finding the ground state can be rather nontrivial.

The next issue is what happens if the temperature is nonzero but still small. The low-$T$ behavior is determined by the nature of the lowest energy states above the ground state, called *elementary excitations*. The elementary excitations are often viewed as particles, since the Bloch theorem guarantees that wavenumber $k$ is a valid quantum number and the $\hbar k$ behaves like the momentum. From the point of view of the thermodynamic properties, each type of elementary excitation is characterized by three properties:

- The $E(k)$ dispersion curve
- Statistical properties (fermions or bosons)
- Fixed number of particles or fixed chemical potential

Further discussion of these topics is given in Chapter 6.

When it comes to finding the elementary excitations, we may try to obtain the complete quantum mechanical solution, or we may obtain valuable insight by emphasizing certain features of the system in question by setting up an appropriate model. For example, when elementary excitations of insulating ferromagnets are considered, it is appropriate to represent the electrons by localized spins. In fact, designing a good *model Hamiltonian* is an art in itself; it must be complex enough to describe a rich set of phenomena, but simple enough to solve.

Many times the model Hamiltonians are represented in terms of particle creation and annihilation operators. The basis for this approach is usually discussed in the textbooks by starting from the harmonic oscillator problem, which is relevant for phonons. The problems in this chapter do not really need an advanced knowledge of *second quantization* (the use of these operators). The advantage of using second quantization is that it takes care of the statistical properties of the particles automatically. Naturally, this advantage is useful only when we have more than one particles in the system, as in Problem 5.1, (b)–(d). A little practice in this method also helps to deal with some of the more complex issues coming in Chapter 9.

Here is a summary of a few basic rules concerning operator algebra. For electrons the operator $a_{i,s}^+$ creates a filled electronic state of spin $s$ on site $i$, and $a_{i,s}$ vacates an electron from the $s, i$ state (these are usually called the *creation and annihilation operators*). In quantum mechanics notation,

$$a_{i,s}^+|0\rangle = |\psi_{i,s}\rangle , \tag{I.5.1}$$

while

$$a_{i,s}|0\rangle = 0 . \tag{I.5.2}$$

Here $|0\rangle$ is the ground state, and $\langle 0|0\rangle = 1$.

For fermions, the *commutation relations* are

$$
\begin{aligned}
a_{i,s}^+ a_{j,s'} + a_{j,s'} a_{i,s}^+ &= \delta_{ss'} \delta_{ij} , \\
a_i a_j + a_j a_i &= 0 , \\
a_i^+ a_j^+ + a_j^+ a_i^+ &= 0 .
\end{aligned}
\tag{I.5.3}
$$

The commutation relations will ensure that no more than one particle can be in any given state. The *particle number operator*, constructed as

$$n_{i,s} = a_{i,s}^+ a_{i,s} , \tag{I.5.4}$$

has eigenvalue of zero or one, depending on the state it is acting upon;

$$n_{i,s} a_{j,s'}^+ |0\rangle = \delta_{ss'} \delta_{ij} |0\rangle \tag{I.5.5}$$

and $n_{i,s}|\psi\rangle = 0$ for any other state.

It is easy to show that linear combinations of the operators, most importantly the plane waves,

$$a_k^+ = \frac{1}{\sqrt{N}} \sum_i e^{ik \cdot R_i} a_i^+$$

$$\text{and} \quad a_k = \frac{1}{\sqrt{N}} \sum_i e^{-ik \cdot R_i} a_i , \tag{I.5.6}$$

satisfy similar commutation relations.

For phonons, or bosons in general, the commutation rules are similar, except that there is a negative sign in the commutators. The phonon state is characterized by wavenumber $q$, and $\epsilon$ may denote, for example, the polarization of the phonon.

$$
\begin{aligned}
b_{q,\epsilon}^+ b_{q',\epsilon'} - b_{q',\epsilon'} b_{q,\epsilon}^+ &= \delta_{\epsilon\epsilon'} \delta_{qq'} , \\
b_q b_{q'} - b_{q'} b_q &= 0 , \\
b_q^+ b_{q'}^+ + b_{q'}^- b_q^+ &= 0 .
\end{aligned}
\tag{I.5.7}
$$

For bosons, states with more than one particle can be created:

$$\begin{aligned}
\alpha_0 b^+ |0\rangle &= |1\rangle \\
\alpha_1 b^+ |1\rangle &= |2\rangle \\
&\vdots \\
\alpha_n b^+ |n\rangle &= |n+1\rangle \, .
\end{aligned} \qquad (I.5.8)$$

For simplicity we dropped the $q$ and $\epsilon$ quantum numbers. The prefactors $\alpha_n$ must be selected so that the $|n+1\rangle$ state is normalized. Starting from $\langle 0|0\rangle = 1$ and using Eqs. I.5.7, one can show from commutation relations that the choice of $\alpha_n = 1/\sqrt{n+1}$ ensures that each $\langle n|n\rangle$ equals 1.

The particle number operator is

$$\mathsf{n}_{q,\epsilon} = b^+_{q,\epsilon} b_{q,\epsilon} \qquad (I.5.9)$$

and its eigenvalue is $n$, the number of particles in the state it is acting on:

$$\mathsf{n}|n\rangle = n|n\rangle \, . \qquad (I.5.10)$$

Finally, for *spin operators* of spin $S$ ($S = 1/2$ for electrons), where the spin components are denoted by $S^x$, $S^y$, $S^z$, it is customary to introduce

$$S^+_i = S^x_i + iS^y_i$$
$$\text{and} \qquad S^-_i = S^x_i - iS^y_i \, . \qquad (I.5.11)$$

The commutators are

$$\begin{aligned}
S^x_i S^y_i - S^y_i S^x_i &= iS^z_i \\
S^y_i S^z_i - S^z_i S^y_i &= iS^x_i \\
S^z_i S^x_i - S^x_i S^z_i &= iS^y_i
\end{aligned} \qquad (I.5.12)$$

or

$$\begin{aligned}
S^+_i S^-_i - S^-_i S^+_i &= 2S^z_i \\
S^-_i S^z_i - S^z_i S^-_i &= S^-_i \\
S^z_i S^+_i - S^+_i S^z_i &= S^+_i \, .
\end{aligned} \qquad (I.5.13)$$

The product $\boldsymbol{S}_i \cdot \boldsymbol{S}_j$ can be expressed as

$$\boldsymbol{S}_i \cdot \boldsymbol{S}_j = S^x_i S^x_j + S^y_i S^y_j + S^z_i S^z_j = \frac{1}{2}(S^+_i S^-_j + S^-_i S^+_j) + S^z_i S^z_j \, . \qquad (I.5.14)$$

In the ground state of a ferromagnet, denoted by $|0\rangle$, all spins point upwards. The $S^-_i$ operator decreases the $z$ component of the spin at site $i$. (For $S = 1/2$ this corresponds to a spin flip: The $z$ component changes from $1/2$ to $-1/2$.) The corresponding new state is denoted by $|i\rangle$; $|i\rangle = S^-_i |0\rangle$. It follows from Eqs. I.5.13 that

$$S_j^z|i\rangle = \begin{cases} (S-1)|i\rangle & \text{for } i \neq j; \\ S|i\rangle & \text{for } i = j. \end{cases}$$

$$S_j^- S_{j+1}^+|i\rangle = \begin{cases} 0 & \text{for } i = j; \\ |i-1\rangle & \text{for } i = j+1; \\ 0 & \text{otherwise}. \end{cases} \qquad (I.5.15)$$

$$S_j^+ S_{j+1}^-|i\rangle = \begin{cases} |i+1\rangle & \text{for } i = j; \\ 0 & \text{for } i = j+1; \\ 0 & \text{otherwise}. \end{cases}$$

## 5.1 Problem: Tight-Binding Model

Electrons propagating in a lattice are often described in terms of electron
creation and annihilation operators. Consider the one-dimensional Hamiltonian

$$H = t \sum_{i,s} a_{i+1,s}^+ a_{i,s} + a_{i,s}^+ a_{i+1,s} , \qquad (I.5.16)$$

where site $i$ is at the position $x_i = ia$ ($a$ is the lattice spacing), and the
operator $a_{i,s}^+$ creates a filled electronic state of spin $s$ in site $i$.

(a) Solve the Schrödinger equation with a one-electron wavefunction,

$$H|\phi(k,s)\rangle = E|\phi(k,s)\rangle . \qquad (I.5.17)$$

(*Hint*: A single, delocalized electron "wave" is described by the wavefunction
$|\phi(k,s)\rangle = N^{-1/2} \sum_p e^{ikx_p} a_{p,s}^+|0\rangle$, where $N$ is the number of sites.)

(b) Show that for two electrons the electron energies add up. The trial
wavefunction is

$$|\phi(k,k',s,s')\rangle = \frac{1}{N} \sum_{p,q} e^{ikx_p} e^{ik'x_q} a_{p,s}^+ a_{q,s'}^+|0\rangle . \qquad (I.5.18)$$

(c) Discuss what happens if $k = k'$ and $s = s'$.

(d) Sometimes it is convenient to define "two-electron excitations" by
considering two electrons with the total $k_{tot} = k + k'$ fixed. Plot the possible
energies of the two-electron excitations as a function of $k_{tot}$.

## 5.2 Problem: Hybridization of Energy Bands

In certain materials a narrow energy band, due to the $d$ orbitals of the atoms,
lies within a broad energy band, originating from $s$-type atomic orbitals. (For
example, in copper the Fermi energy is in the middle of the broad band and
the narrow band lies somewhat below $E_F$; in nickel $E_F$ cuts into the narrow

band.) A simple one-dimensional model can be constructed from a tight-binding Hamiltonian for the broad band, $H_c = t \sum_i (c_i^+ c_{i+1} + c_i^- c_{i+1}^+)$, and from $H_d = E_d \sum_i d_i^+ d_i$, to describe nonoverlapping, localized electrons. Here $c_i^+$ and $c_i$ create and annihilate an $s$-wave-type electron in unit cell $i$; $d_i^+$ and $d_i$ create and annihilate a localized electron. The electron operators obey the usual anticommutation relations. The total Hamiltonian includes a term to describe the possible hopping of the electron from the localized orbit to the delocalized state and back:

$$H_c = t \sum_i (c_i^+ c_{i+1} + c_i c_{i+1}^+) + E_0 \sum_i d_i^+ d_i + \sum_i \Delta(c_i^+ d_i + c_i d_i^+) \quad (I.5.19)$$

The spin indices are not used, since none of the processes involve spin flips.

   Calculate the energy spectrum of the of the electrons for (a) $\Delta = 0$ and (b) for $\Delta \neq 0$.


## 5.3 Problem: Polarons

A "polaron" is formed when an electron has a bound state in the potential created by a distorted lattice, and the lattice distortion is due to the electron itself. To elucidate the microscopic mechanism of polaron formation, consider a single electron in one dimension. The crystal lattice is represented by a continuous medium, and the potential felt by the electrons is $V = \lambda du/dx$, where $u(x)$ is the displacement of the atoms from their original position and $\lambda$ is the electron–phonon coupling constant. The energy density associated with the deformation is $1/2B(du/dx)^2$, where $B$ is the bulk modulus. Show that in one dimension the coupled electron–phonon system is always unstable against the formation of polarons. Calculate the size of the polaron. (Assume that the deformation is in the form of $u = u_0 \tanh x/a_0$, where $u_0$ and $a_0$ are variable parameters; $a_0$ is the size of the polaron.)


## 5.4 Problem: Polaritons

Determine the frequencies of the longitudinal and the transverse waves in a polar insulator or semiconductor, where the $\Gamma \to 0$ limit of the oscillator model, Eq. I.8.18, describes the dielectric function. Discuss the wavenumber ($k$) dependence of the transverse waves. Show that the *Lyddane–Sachs–Teller relation*,

$$\frac{\omega_L^2}{\omega_T^2} = \frac{\epsilon_1(\omega \to 0)}{\epsilon_1(\omega \to \infty)}, \quad (I.5.20)$$

holds, where $\epsilon_1$ is the real part of the dielectric function, $\omega_L$ is the longitudinal wave frequency, and $\omega_T$ is the transverse wave frequency at $k \gg \omega/c$.

## 5.5 Problem: Excitons

Excitons are bound electron–hole pairs, typically observed in semiconductors. Assume that the conduction and valence bands are described by

$$E = \pm\sqrt{\frac{\hbar^2 k^2}{m} + \Delta^2} \qquad (I.5.21)$$

with $m$ being the free-electron mass and $\Delta = 1$ eV. Estimate the binding energy of an exciton, assuming a dielectric constant of $\varepsilon = 16$. At what temperatures could we observe excitons?

## 5.6 Problem: Holstein–Primakoff Transformation

Show that the operators defined by the Holstein–Primakoff transformation,

$$
\begin{aligned}
S^+ &= \hbar\sqrt{2S}a^+\sqrt{1 - \frac{a^+ a}{2S}} \;, \\
S^- &= \hbar\sqrt{2S}\sqrt{1 - \frac{a^+ a}{2S}}\,a \;, \qquad (I.5.22) \\
S^z &= \hbar(a^+ a - S) \;,
\end{aligned}
$$

satisfy the spin commutation relations. Here $a^+$ and $a$ are boson creation and annihilation operators, and $S$ is the magnitude of the spin.[1]

## 5.7 Problem: Dyson–Maleev Representation

In the Dyson–Maleev representation the spin operators are expressed by the boson creation and annihilation operators $a^+$ and $a$ as

$$
\begin{aligned}
S^+ &= \hbar\sqrt{2S}a^+\left(1 - \frac{a^+ a}{2S}\right) \;, \\
S^- &= \hbar\sqrt{2S}a \;, \qquad (I.5.23) \\
S^z &= \hbar(a^+ a - S) \;,
\end{aligned}
$$

where $S$ is the magnitude of the spin. Show that the operators satisfy the spin commutation relations.

---

[1] The Holstein–Primakoff transformation is discussed, for example, by Mattis [8] p. 154.

## 5.8 Problem: Spin Waves

Determine the magnon (spin wave) dispersion curve in the one-dimensional ferromagnetic Heisenberg model,

$$H = -J \sum_i S_i S_{i+1} \qquad (1.5.24)$$

(where $J > 0$ and $S = (S^x, S^y, S^z)$ is a spin operator), and in the Ising model, $H = -J \sum_i S_i S_{i+1}$ (where $S \equiv S^z = \pm 1/2$). Site $i$ is at position $R_i = ai$, where $a$ is the lattice spacing.

## 5.9 Problem: Spin Waves Again

Derive the Heisenberg Hamiltonian in the Dyson–Maleev representation of spins. Neglect the terms containing more than two boson operators, and determine the elementary excitations of the system. Under what circumstances is this approximation justified? (The Heisenberg Hamiltonian is $H = -\sum_{ij} J_{ij} S_i S_j = -J \sum_i S_i S_{i+1}$; the Dyson–Maleev representation expresses the spin operator at site $i$ as $S_i^+ = \sqrt{2S} a_i^+ \left(1 - \frac{a_i^+ a_i}{2S}\right)$, $S_i^- = \sqrt{2S} a_i$ and $S_i^z = a_i^+ a_i - S$, where $a_i^+$ and $a_i$ are boson creation and annihilation operators, in units of $\hbar = 1$.)[2]

## 5.10 Problem: Anisotropic Heisenberg Model

The anisotropic Heisenberg model is defined by the Hamiltonian

$$H = -\sum_{i,j} \left[ J'_{ij} S_i^z S_j^z + \frac{1}{2} J_{ij} (S_i^+ S_j^- + S_i^- S_j^+) \right] . \qquad (1.5.25)$$

Consider a one-dimensional system with nearest-neighbor interactions.

$$H = -\sum_i \left[ J' S_i^z S_{i+1}^z + \frac{1}{2} J (S_i^+ S_{i+1}^- + S_i^- S_{i+1}^+) \right] , \qquad (1.5.26)$$

and ferromagnetic coupling, $J > 0$, $J' > 0$.

(a) Show that for $J' > J$ the magnon wavefunctions are similar to the isotropic Heisenberg model solutions, but they have a nonzero energy at $k = 0$.

(b) Show that for $J' < J$ the ferromagnetically ordered state (with magnetization aligned in the $z$ direction) is *not* the ground state of the Hamiltonian.

---

[2] See Mattis [8] p. 76.

## 5.11 Problem: Solitons

Some nonlinear partial differential equations have solitary wave, or *soliton* solutions. Solitons have a finite "extension" and they "survive" collisions with other solitons. These concepts are well illustrated by the sine–Gordon equation:

$$\frac{\partial^2 \varphi}{\partial x^2} - \frac{\partial^2 \varphi}{\partial t^2} = m \sin \varphi \,, \tag{I.5.27}$$

where $\varphi$ is a variable depending on the position $x$ and time $t$, and $m$ is a parameter. The simplest physical system represented by this equation is an array of pendula. Electronic excitations in polymers like $(CH)_x$, magnetic chains, and other systems are also modeled by this equation.

(a) Show that the "single-soliton",

$$\varphi = \pm 4 \tan^{-1} \exp(\pm m\gamma(x - vt) + \delta) \tag{I.5.28}$$

is a solution to Eq. I.5.27. Calculate $\gamma$ in terms of $v$ and $\delta$. Plot the solution for $v = 0$. Show that $\varphi$ is close to 0 or $2\pi$ for almost every $x$, except in a range of width $\Delta$ and estimate $\Delta$ for $v = 0$. Calculate the energy $E_0$ stored in the system, relative to the energy of the $\varphi = 0$ solution.

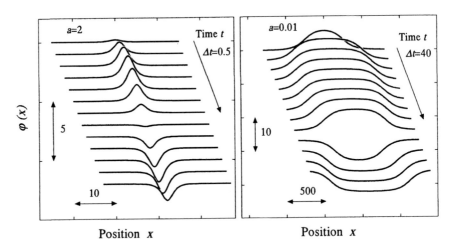

Fig. I.5.1. Solution of Eq. I.5.27 as expressed in Eq. I.5.29 for $a = 2$ and $a = 0.01$ and $m = 1$.

(b) Show that the "two-soliton" (or "breather"),

$$\varphi = 4 \tan^{-1} \frac{1}{a} \frac{\sin(m\gamma a t)}{\cosh(m\gamma x)} \,, \tag{I.5.29}$$

is also a solution. Calculate $\gamma$. Calculate the energy and show that it is always less than $2E_0$.

(c) In Figure I.5.1 the two-soliton solution is plotted for two values of $m$. Demonstrate that, for small $a$ and for most values of the time $t$, the solution can be approximated by the sum of two, appropriately selected "one-soliton" solutions.

# 6 Thermodynamics of Noninteracting Quasiparticles

The problems in this chapter deal with the thermal properties of materials; some of them could equally well fit into a book on statistical mechanics or thermodynamics. These topics do, however, come up often when studying condensed matter physics and we therefore included a brief introduction and set of problems. Nevertheless, we give only a very short and rudimentary summary of the ideas and concepts used in these areas. For a more thorough treatment of these and related topics, we refer the reader to statistical mechanics and thermodynamics textbooks such as Reif [17], Reichl [18] or many others.

According to the first law of thermodynamics, the energy change $dE$ of a system is expressed as

$$dE = TdS - pdV + \mu dN + \ldots \, , \qquad (\text{I.6.1})$$

where $TdS = \Delta Q$ is the heat transfer, and the other terms describe various other ways of transferring energy between the system and its surrounding. We refer the reader to textbooks on thermodynamics for the discussion of *intensive variables*, like temperature $(T)$, pressure $(p)$, and chemical potential $(\mu)$, and *extensive variables*, like entropy $(S)$, volume $(V)$, and number of particles $(N)$. Depending on the nature of the system, there may be other similar terms in this expression; for example, in magnetic systems, there is an $HdM$ term (here $H$ is the external magnetic field and $M$ is the magnetization). The function $E = E(S, V, N, ...)$ fully characterizes the thermodynamic behavior of a system. The equations of state are obtained by calculating the partial derivatives of this function:

$$T = \left(\frac{\partial E}{\partial S}\right)_{V,N}, \qquad -p = \left(\frac{\partial E}{\partial V}\right)_{S,N}, \qquad \mu = \left(\frac{\partial E}{\partial N}\right)_{S,V}. \qquad (\text{I.6.2})$$

For these equations to work, the energy must be a function of the extensive variables (like $S$, $V$, $N$). In order to eliminate any ambiguity, the variables kept constant are indicated in the subscript of the partial derivatives.

In many cases the problem is better formulated in terms of a mix of intensive and extensive variables. For this purpose the other *thermodynamic potentials* were introduced. They are obtained from the energy function by a

53

mathematical procedure called the *Legendre transformation*. Here are a few examples:

$$
\begin{aligned}
F(T,V,N) &= E - TS & &\text{Helmholtz potential (or free energy)}\\
H(S,p,N) &= E + pV & &\text{Enthalpy}\\
G(T,p,N) &= E - TS - pV & &\text{Gibbs potential}\\
U(T,V,\mu) &= E - TS - \mu N & &\text{Grand canonical potential.}
\end{aligned}
\tag{I.6.3}
$$

The "natural variables" of each of the thermodynamic potentials are different (as indicated above). The purpose of the Legendre transformation is to make sure that relations analogous to Eqs. I.6.1 still hold:

$$
\begin{aligned}
dF &= -SdT - pdV + \mu dN + \dots\,,\\
dH &= TdS + Vdp + \mu dN + \dots\,,\\
dG &= -SdT + Vdp + \mu dN + \dots\,,\\
dU &= -SdT - pdV - Nd\mu + \dots\,.
\end{aligned}
\tag{I.6.4}
$$

Consequently, relations analogous to Eqs. I.6.2 also hold. For example:

$$
-S = \left(\frac{\partial F}{\partial T}\right)_{V,N}\,;\quad
-p = \left(\frac{\partial F}{\partial V}\right)_{T,N}\,;\quad
\mu = \left(\frac{\partial F}{\partial N}\right)_{T,V}\,.
\tag{I.6.5}
$$

The usefulness of this mathematical construction becomes clear when experimentally measured quantities are considered. The general definitions are as follows:

$$
\begin{aligned}
c &= \frac{1}{N}\frac{\Delta Q}{\Delta T} & &\text{Specific heat}\\
\kappa &= -\frac{1}{V}\frac{\Delta V}{\Delta p} & &\text{Compressibility}\\
B &= 1/\kappa & &\text{Bulk modulus}\\
\alpha &= \frac{1}{V}\frac{\Delta V}{\Delta T} & &\text{Coefficient of thermal expansion.}
\end{aligned}
\tag{I.6.6}
$$

Usually each of these quantities is measured so that some variables are kept constant during the process; if ambiguity exists, then these variables are listed in the subscript. Using $TdS = \Delta Q$ and the definitions in Eqs. I.6.6 and Eqs. I.6.4, one obtains

$$
c_V \equiv c_{V,N} = \frac{1}{N}T\left(\frac{\partial S}{\partial T}\right)_{V,N} = -\frac{T}{N}\left(\frac{\partial^2 F}{\partial T^2}\right)_{V,N}
\tag{I.6.7}
$$

$$
c_p \equiv c_{p,N} = \frac{1}{N}T\left(\frac{\partial S}{\partial T}\right)_{p,N} = -\frac{T}{N}\left(\frac{\partial^2 G}{\partial T^2}\right)_{p,N}
\tag{I.6.8}
$$

$$
\kappa_T = -\frac{1}{V}\left(\frac{\partial V}{\partial p}\right)_{T,N} = -\frac{1}{V}\left(\frac{\partial^2 G}{\partial p^2}\right)_{T,N}
\tag{I.6.9}
$$

$$B_T = \frac{1}{\kappa_T} = -V\left(\frac{\partial p}{\partial V}\right)_{T,N} = V\left(\frac{\partial^2 F}{\partial V^2}\right)_{T,N} \quad \text{(I.6.10)}$$

$$\alpha = \frac{1}{V}\left(\frac{\partial V}{\partial T}\right)_N = \frac{1}{V}\left(\frac{\partial^2 G}{\partial T \partial p}\right)_N . \quad \text{(I.6.11)}$$

Once the mathematical framework is established, it is easy to find relationships between the various "second derivatives". For example,

$$B_S = \frac{c_p}{c_V \kappa_T} ,$$

$$c_V = c_p - \frac{TV\alpha^2}{N\kappa_T} , \quad \text{(I.6.12)}$$

where $B_S$ is the adiabatic bulk modulus. These relationships may be used to answer certain questions even if the full equation of state of a material is not known, yet some of the material properties (like specific heat, compressibility, etc.) are given.

For a system made of a large number of identical particles, with known rules governing the motion of these components, the methods of statistical mechanics are used to calculate the fundamental equations of state. The goal of the calculations is to determine one of the two *partition functions* $(Z)$; each of these, in turn, is related to the appropriate thermodynamic potential. For systems of fixed number of particles, the *canonical partition function* is used and the free energy is obtained:

$$Z_N = \sum_{\alpha:\text{states}} e^{-E_\alpha/k_B T} , \quad F(T,V,N) = -k_B T \log Z_N . \quad \text{(I.6.13)}$$

If the number of particles is not fixed, the *grand canonical partition function* is used:

$$Z = \sum_{\alpha:\text{states}} e^{-(E_\alpha - \mu)/k_B T} , \quad U(T,V,\mu) = -k_B T \log Z . \quad \text{(I.6.14)}$$

Note that, as long as the interactions are short ranged, the partition function of a composite system made of two independent systems is the product of the partition functions of the components: $Z_{a+b} = Z_a Z_b$. This feature ensures that the thermodynamic potentials are additive.

Hidden behind these formulae is some simple physics. For example, the total energy of the system can be expressed in terms of the canonical partition function as

$$\begin{aligned} E &= F + TS = F - T\frac{\partial F}{\partial T} = -k_B T \log Z_N + k_B T\frac{\partial \log T Z_N}{\partial T} \\ &= -k_B T \log Z_N + k_B T \log Z_N + k_B T^2 \frac{\partial \log Z_N}{\partial T} \end{aligned}$$

$$= \frac{1}{Z_N} \sum_{\alpha:\text{states}} E_\alpha e^{-E_\alpha/k_B T}$$

$$= \sum_{\alpha:\text{states}} E_\alpha p_\alpha , \tag{I.6.15}$$

where $p_\alpha = e^{-E_\alpha/k_B T}/Z_N$ is the probability of finding the system at energy $E_\alpha$.

In other respects, the apparent simplicity of the equations is somewhat misleading. For classical systems of $N$ particles, the $\sum_{\text{states}}$ is in fact an integral over the $6N$-dimensional phase space; in quantum systems, we must know the full energy spectrum (including the degeneracy for each energy) or the diagonal of the matrix elements of the operator $e^{-H/k_B T}$ with a complete $N$ particle basis set.[1] For systems of interacting particles there is no easy and general way to calculate the partition function. The vast majority of the interesting problems have no exact solution.

A major simplification can be obtained for systems of noninteracting particles with known single-particle energy spectrum: The energy of a component particle[2] in state $|i\rangle$ is $\mathcal{E}_i$. A rather elaborate calculation leads to the partition functions

$$Z = \prod_i \exp\{\exp\frac{-(\mathcal{E}_i - \mu)}{k_B T}\} \qquad \text{Classical} \qquad \text{(I.6.16)}$$

$$Z = \prod_i \frac{1}{1 + \exp\left[\frac{-(\mathcal{E}_i - \mu)}{k_B T}\right]} \qquad \text{Bose–Einstein} \qquad \text{(I.6.17)}$$

$$Z = \prod_i \frac{1}{1 - \exp\left[\frac{-(\mathcal{E}_i - \mu)}{k_B T}\right]} \qquad \text{Fermi–Dirac} . \qquad \text{(I.6.18)}$$

Let us calculate the total energy of a system. Similar to Eq. I.6.15, but now using grand canonical partition function, we get

$$\begin{aligned} E &= U + TS + \mu N = U - T\frac{\partial U}{\partial T} - \mu \frac{\partial U}{\partial \mu} \\ &= -k_B T \log Z + k_B T \frac{\partial T \log Z}{\partial T} + k_B T \mu \frac{\partial \log Z}{\partial \mu} \\ &= \sum_i k_B T^2 \frac{\partial \log Z_i}{\partial T} + k_B T \mu \frac{\partial \log Z_i}{\partial \mu} \end{aligned} \tag{I.6.19}$$

---

[1] For quantum systems, the most general definition of the partition function is $Z_N = \text{Tr}\{e^{-H/k_B T}\}$, where H is the Hamiltonian operator. The trace is invariant of the choice of basis set; for the basis set where the Hamiltonian is diagonal, this equation is equivalent to the definition given above.

[2] In this chapter, as well as in other cases when ambiguity may exist, we adopt the convention that the single-particle energies will be denoted by $\mathcal{E}$; the energy of the system is $E$.

$$= \sum_i (\mathcal{E}_i - \mu)\, f(\mathcal{E}_i)\,.$$

Furthermore, the total number of particles is

$$N = -\frac{\partial U}{\partial \mu} = \sum_i f(\mathcal{E}_i)\,, \tag{I.6.20}$$

where the function $f(\mathcal{E}_i) \equiv f(\mathcal{E})$ is

$$f_{cl}(\mathcal{E}) = e^{-(\mathcal{E}-\mu)/k_B T} \qquad \text{Classical} \tag{I.6.21}$$

$$f_{BE}(\mathcal{E}) = \frac{1}{e^{(\mathcal{E}-\mu)/k_B T} - 1} \qquad \text{Bose–Einstein} \tag{I.6.22}$$

$$f_{FD}(\mathcal{E}) = \frac{1}{e^{(\mathcal{E}-\mu)/k_B T} + 1} \qquad \text{Fermi–Dirac} \tag{I.6.23}$$

The result presented in Eq. I.6.19 resembles Eq. I.6.15 in that the total energy is expressed as the weighted average of the energies, with temperature-dependent weight factors. (The chemical potential $\mu$ does not appear in Eq. I.6.15, because the canonical partition function was used.) Note the tremendous progress we have made: Instead of trying to perform a sum over all possible states of the entire system, we now have Eqs. I.6.19 and I.6.20 with the sum over the single-particle states. Naturally, this is only valid for non-interacting particles.

An inspection of Eqs. I.6.19 and I.6.20 suggests a simple physical interpretation: The function $f(\mathcal{E})$ tells us the expected average number of particles at a given energy level and the total energy is the weighted sum of the single-particle energies. As discussed in Chapter 4, introducing the density of states further simplifies the calculations:

$$E = V \int d\mathcal{E} \; \mathcal{E} f(\mathcal{E}) g(\mathcal{E}) \tag{I.6.24}$$

$$N = V \int d\mathcal{E} \; g(\mathcal{E}) f(\mathcal{E})\,. \tag{I.6.25}$$

Setting the particle number to a given value determines the chemical potential. Figure I.6.1 illustrates that imposing this condition will have a profound effect on the energy states occupied by the particles. Fermions have a maximum occupancy of 1, whereas bosons overwhelmingly prefer the low-energy states.[3]

The simplicity of the results represented by Eqs. I.6.24 and I.6.25 , combined with the concept of elementary excitations (as discussed in Chapter 5), leads to a commonly employed method of determining the thermodynamic properties of a system of interacting particles. The first, and most difficult,

---

[3] In fact, with a fixed number of particles and at high enough densities the ideal (noninteracting) gas of bosons exhibits Bose–Einstein condensation, when *all* particles are in the ground state.

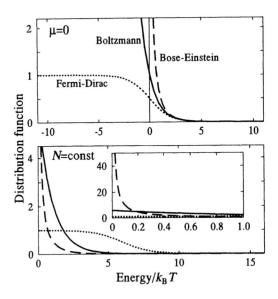

**Fig. I.6.1.** The $f(E)$ distribution functions plotted as a function of energy. The upper panel is a representation of Eqs. I.6.21– I.6.23, with chemical potential $\mu = 0$. The plot in the lower panel corresponds to the number of particles being fixed, with the assumption that the density of states is zero for $E < 0$ and constant up to energies much higher than the thermal energy, $k_B T$. The insert shows the lowest energy part, where the Bose–Einstein distribution diverges.

step is the determination of the spectrum of the elementary excitations and the corresponding density of states. Except for exotic (and therefore very interesting) systems, the elementary excitations are often either bosons or fermions; we can also specify if the number of particles is fixed or variable. For example, in crystalline insulators the elementary excitations are bosons (phonons) and the number of particles is arbitrary. Similarly, in a ferromagnet, the elementary excitations are bosons (magnons) with no restriction on the particle number.[4] For electrons in a metal the elementary excitations are fermions, with fixed number of particles.

Once these properties are known, the calculation of the thermodynamic properties becomes simple: Insert the proper density of states and distribution function into Eqs. I.6.24 and I.6.25, and use Eq. I.6.25 either to calculate the chemical potential (if the number of particles is fixed) or to calculate the number of particles (by inserting $\mu = 0$, if the number of particles is variable).

One has to keep in mind that from the perspective of thermodynamic potentials the results of these calculations are often in a "nonstandard" form: The energy should be expressed as $E = E(S, V, N)$, whereas in Eq. I.6.24 we obtain $E = E(T, V, \mu)$. In principle, it is relatively easy to convert $E(T, V, \mu)$ to $E(S, V, N)$ or to calculate any of the other thermodynamic potentials. First, for a system with fixed number of particles (like electrons in a metal) the chemical potential can be expressed by using Eq. I.6.25, and this way we can obtain $E = E(T, V, \mu(T, V, N)) \equiv E(T, V, N)$. Second, at constant volume and particle number the first law, Eq. I.6.1 reduces to $dE = TdS$. Therefore $dS = 1/T dE$ and

---

[4] However, the excitation spectrum of phonons and magnons are different, resulting in different density of states.

$$S(T,V,N) = \int_{T_0}^{T} \frac{1}{T'} \frac{\partial E(T',V,N)}{\partial T'} \, dT' \ . \tag{I.6.26}$$

The lower limit of integration must be chosen so that the *third law of thermodynamics* is satisfied and the entropy is zero at $T = 0$. The entropy function can be inverted to obtain $T = T(S,V,N)$, and the result can be used to express the temperature term in $E = E(T,V,N) = E(T(S,V,N),V,N) \equiv E(S,V,N)$. In practice the calculation often leads to integrals which cannot be obtained in closed form and for which the inversion is impossible to perform. It is somewhat more hopeful to obtain the free energy in terms of its appropriate variables:

$$
\begin{aligned}
F(T,V,N) &= E - TS = E - T \left( \int_{T_0}^{T} \frac{1}{T'} \frac{\partial E(T',V,N)}{\partial T'} \, dT' \right) \\
&= E - T \left( \frac{E}{T} \Big|_{T_0}^{T} + \int_{T_0}^{T} \frac{1}{T'^2} E(T',V,N) \, dT' \right) \\
&= -T \int_{T_0}^{T} \frac{1}{T'^2} E(T',V,N) \, dT' + \frac{T}{T_0} E(T_0).
\end{aligned}
\tag{I.6.27}
$$

For the calculation of the specific heat at constant volume and particle number, the whole process can be simplified. According to Eq. I.6.1, under those conditions $dE = TdS = \Delta Q$ and therefore Eq. I.6.7 becomes

$$c_{V,N} = \frac{1}{N} T \left( \frac{\partial S}{\partial T} \right)_{V,N} = \frac{1}{N} \left( \frac{\partial E}{\partial T} \right)_{V,N} . \tag{I.6.28}$$

At zero temperature (when the $TdS$ term is zero) another useful relationship can be obtained for the bulk modulus by using $dE = -pdV$:

$$B_{T=0,N} = -V \frac{\partial p}{\partial V} = V \frac{\partial^2 E}{\partial V^2} \ . \tag{I.6.29}$$

For fermions of fixed particle number, at temperatures much less than the *Fermi temperature* $(T_F = E_F/k_B)$, the Bethe–Sommerfeld expansion provides a way to handle the mathematical difficulties encountered in calculating the integrals I.6.24 and I.6.25:

$$E = E(T,V,N) = E_0(V) + \frac{\pi^2}{6} V (k_B T)^2 g(E_F) \tag{I.6.30}$$

$$\mu = E_F - \frac{\pi^2}{6} (k_B T)^2 \frac{g'(E_F)}{g(E_F)}, \tag{I.6.31}$$

where $g'(E) = dg(E)/dE$.

In this approximation, the specific heat contribution of electrons in a metal is

$$c_V = \frac{\pi^2}{3} k_B^2 T g(E_F) \ . \tag{I.6.32}$$

## 6.1 Problem: Specific Heat of Metals and Insulators

Sketch the temperature dependence of the specific heat contribution of a conduction electrons and lattice vibrations for temperatures ranging from 0 to 500 K. If possible, scale your plot in terms of these parameters: $E_F$ (Fermi energy), $n$ (electron density), $\omega_D$ (Debye frequency), and $N$ (number density of atoms). Also indicate the characteristic temperature dependence for the various regimes. Estimate the ratio of the two contributions to the specific heat at $T = 1K$ and at $T = 500$ K, assuming typical values for the above parameters.

## 6.2 Problem: Number of Phonons

Calculate the total number of phonons in the Debye approximation. Discuss the low and high temperature limits – that is, $k_B T \ll \hbar c_s k_D$ and $k_B T \gg \hbar c_s k_D$. Here $k_D$ is the Debye wavenumber, $c_s$ is the sound velocity and $s = 1, 2$, and 3 correspond to the three phonon modes. The number of atoms in the system is $N_a$.

## 6.3 Problem: Energy of the Phonon Gas

In the Debye approximation, calculate the total phonon energy $E$ and the specific heat $c_V$ in the low- and high-temperature limits.

## 6.4 Problem: Bulk Modulus of Phonon Gas

The volume dependence of the phonon mode frequencies is approximated by the Grüneisen relation, $\Delta\omega(k)/\omega(k) = -\gamma \Delta V/V$, where $\gamma$ is a constant of the order of unity. Calculate the pressure and the bulk modulus of the phonon gas. Use the Debye approximation to estimate the order of magnitude of this quantity and compare it to the measured bulk moduli of solids.

## 6.5 Problem: Phonons in One Dimension

What are the characteristic low-$T$ and high-$T$ temperature dependencies of the number of phonons $(N)$, phonon energy $(E)$, and specific heat $(c_V)$ for a one-dimensional phonon gas, if the energy dispersion is described in the Debye approximation?

## 6.6 Problem: Electron–Hole Symmetry

We have a semiconductor with two bands: The lower band is described by $E(\boldsymbol{k})$, the upper band by $E_0 - E(\boldsymbol{k})$. The two bands are separated by an energy gap of $2\Delta$.

(a) Calculate the chemical potential as a function of the temperature.
(b) Show that the system can be treated using classical statistics at low temperatures.

## 6.7 Problem: Entropy of the Noninteracting Electron Gas

According to Eqs. I.6.24 and I.6.25 the energy and the particle number of the free-electron gas can be expressed as simple integrals over the density of states and Fermi functions:

$$E \;=\; \int d\mathcal{E}\; \mathcal{E} f(\mathcal{E}) g(\mathcal{E}) \qquad\qquad (\text{I.6.33})$$

$$N \;=\; \int d\mathcal{E}\; g(\mathcal{E}) f(\mathcal{E}) \qquad\qquad (\text{I.6.34})$$

Express the entropy of the free-electron gas similarly in the form of

$$S = \int g(\mathcal{E}) \mathcal{F}(f)\, d\mathcal{E} \;, \qquad\qquad (\text{I.6.35})$$

where $\mathcal{F}(f)$ is a function of the Fermi function $f(T, \mu)$.

## 6.8 Problem: Free Energy with Gap at the Fermi Energy

Consider a "generic" density of states which describes an energy gap in the spectrum of electrons:

$$g(\mathcal{E}) = g_0 \frac{|\mathcal{E} - \mathcal{E}_{\mathrm{F}}|}{\sqrt{(\mathcal{E} - \mathcal{E}_{\mathrm{F}})^2 - \Delta^2}} \;, \qquad\qquad (\text{I.6.36})$$

where the gap between the lower and upper band is $2\Delta$, and the density of states of the unperturbed system can be approximated by a constant, $g_0 = 1/VW$, where $V$ is the volume of the system. The gap opens at the Fermi energy $\mathcal{E}_{\mathrm{F}}$.[5]

---

[5] This density of states is encountered in the BCS theory of superconductivity (see Tinkham [20] pp. 41–44), or in quasi-one-dimensional charge density wave systems (see Problem 4.9).

(a) Calculate, at zero temperature, the leading term in the change of the total energy of the electrons as $\Delta$ is increased from zero to a finite value.

(b) Investigate the contribution of the $TS$ term in the free energy, $F = E - TS$, at finite but small ($k_B T \ll \Delta$) temperatures.

## 6.9 Problem: Bulk Modulus at $T = 0$

Calculate the bulk modulus of the free electron gas at zero temperature.

## 6.10 Problem: Temperature Dependence of the Bulk Modulus

Calculate the leading term in the temperature dependence of the bulk modulus of the three-dimensional free-electron gas.

## 6.11 Problem: Chemical Potential of the Free-Electron Gas

Calculate the leading term in the temperature dependence of the chemical potential for one-, two-, and three-dimensional free-electron gases.

## 6.12 Problem: EuO Specific Heat

The specific heat of EuO at low temperatures is proportional to $T^{3/2}$. Is EuO a metal or an insulator? What is a possible simple spectrum $E(k)$ giving rise to this specific heat? What type of elementary excitation does this spectrum have?

## 6.13 Problem: Magnetization at Low Temperatures

Determine the number of magnons in a ferromagnet at temperatures much lower than its transition temperature. Assume a three-dimensional, simple cubic system described by the Heisenberg Hamiltonian with nearest-neighbor coupling $J$. Use the result to calculate the temperature dependence of the magnetization.

## 6.14 Problem: Electronic Specific Heat

In the tight-binding approximation the band structure of electrons is described by

$$E = -(E_1 \cos k_x a + E_2 \cos k_y b + E_3 \cos k_z c) \qquad (I.6.37)$$

Show that for nearly empty or nearly full band the electronic specific heat is equivalent to the free-electron specific heat with effective mass of $m' = |\det M|^{1/3}$, where M is the effective mass tensor.


## 6.15 Problem: Quantum Hall Effect

An electron is confined to a plane. Within the plane it moves freely, except the motion is restricted to an area of $A = L^2$. A magnetic field $B$ is applied perpendicular to the plane.

(a) Calculate the energy levels.
(b) Calculate the degeneracy of each energy level.
(c) Assume that we have $N = nA$ electrons and that the interaction between the electrons is neglected. Calculate and plot the magnetic field dependence of the chemical potential $\mu$ for temperatures $k_B T \ll \hbar \omega_c$, where $\omega_c = Be/mc$ is the cyclotron frequency. (The peculiar behavior of this system is related to the *integer quantum Hall effect* as discussed by Ibach and Lüth [4] p. 322.)
(d) Electrons of density $n' = 10^{18}$ cm$^{-3}$ are confined into a 100 Åthick layer. What is the typical magnetic field required to see the effect discussed above? What temperatures are low enough at this field?

# 7 Transport Properties

In investigating thermodynamic properties like specific heat, magnetic susceptibility, or the thermal expansion coefficient, an implicit assumption is made that the system is in quasistatic equilibrium: The change of externally controlled parameters (temperature, magnetic field, etc.) is very slow, and the processes are reversible. However, the time evolution of the system, or the behavior under nonequilibrium conditions, is of great importance as well. This is the subject of *irreversible thermodynamics*, which is related to the corresponding microscopic *transport theory*.

In this chapter we will concentrate on systems where the deviation from equilibrium is small. When the external perturbation (the temperature gradient, electric potential gradient ...) is switched off, the system returns to its equilibrium state in the time scale set by the *relaxation time*. If the perturbation is maintained over a time scale much longer than the relaxation time then a steady flow of energy, particles, charge, and so forth, develops. For small perturbations the response can be characterized by the proportionality constants (called *transport coefficients*) between these quantities.

Here are the definitions for a few commonly used transport coefficients. Note that while one of the parameters of the system is modified (*e.g.*, a temperature gradient is imposed), other, well-specified parameters must be kept under control (*e.g.*, no electric current is allowed):

$$
\begin{array}{llll}
-\text{Thermal conductivity}: & \kappa = -j_Q/\nabla T, & j_e = 0 \\
-\text{Electrical conductivity}: & \sigma = j_e/E, & \nabla T = 0 \\
-\text{Peltier coefficient}: & \Pi = j_Q/j_e, & \nabla T = 0 \\
-\text{Thermopower}: & S = E/\nabla T, & j_e = 0 \\
-\text{Hall coefficient}: & R_\mathrm{H} = E_y/(Hj_x), & j_y = 0, \nabla T = 0
\end{array}
$$

where $j_Q, j_e, \nabla T, E$, and $H$ are the heat current, electric current, temperature gradient, electric field, and magnetic field, respectively.[1]

Due to the time-reversible nature of microscopic processes, the transport coefficients are not entirely independent. The *Onsager relations* can be used

---

[1] To illustrate the concept, while keeping the formalism simple, a one-dimensional flow was assumed for the first four items; in the definition of the Hall coefficient the current density, electric field, and magnetic field point in the $x$, $y$, and $z$ directions of a Cartesian coordinate system, respectively.

to calculate some coefficients if some others are known. The most commonly used relation connects electric and heat transports; it can be best formulated in terms of the $L_{pq}$ *generalized transport coefficients*.[2] In zero magnetic field we have

$$
\begin{aligned}
j_e &= L_{11}E - L_{12}\frac{\nabla T}{T} \qquad \text{Electric current} \\
j_Q &= L_{21}E - L_{22}\frac{\nabla T}{T} \qquad \text{Thermal current.}
\end{aligned} \tag{I.7.1}
$$

The Onsager relation turns out to be a symmetry condition on the $L_{pq}$ coefficients:

$$
L_{21} = L_{12} \tag{I.7.2}
$$

Using the definition of the transport coefficients, along with Eqs. I.7.1 and I.7.2, straightforward calculations result in

$$
\begin{aligned}
\sigma &= L_{11} \\
\kappa &= \left(-\frac{L_{12}^2}{L_{11}} + L_{22}\right)\frac{1}{T} \\
\Pi &= L_{12}/L_{11} \\
S &= \frac{L_{12}}{TL_{11}}.
\end{aligned} \tag{I.7.3}
$$

For simplicity, we assumed that the system is isotropic, and we replaced the L tensors by $L$ scalars.

In a multicomponent system, or in the presence of a magnetic field, many other transport coefficients can be defined. Here we will discuss only the Hall effect. If a magnetic field $H$ points in the $z$ direction of a Cartesian coordinate system, and the electric current is constrained to be in the $xy$ plane, then for an isotropic sample with no temperature gradient, the relationship between the electric field and the electric current is

$$
\begin{aligned}
E_x &= \rho j_x - HR_H j_y \\
E_y &= HR_H j_x + \rho j_y,
\end{aligned} \tag{I.7.4}
$$

where $j$ is the electric current, and $\rho$ is the resistivity.

Inverting Eq. I.7.4 yields

$$
\begin{aligned}
j_x &= \frac{\sigma}{(\sigma R_H H)^2 + 1}E_x + \frac{\sigma^2 R_H H}{(\sigma R_H H)^2 + 1}E_y \\
j_y &= -\frac{\sigma^2 R_H H}{(\sigma R_H H)^2 + 1}E_x + \frac{\sigma}{(\sigma R_H H)^2 + 1}E_y.
\end{aligned} \tag{I.7.5}
$$

In this context the tensor in the equation $E = \hat{\rho}(H)j$ is called the *Hall resistivity* tensor and $\hat{\sigma} = \hat{\rho}^{-1}$ is the *Hall conductivity* tensor. The antisymmetric

---

[2] For each $p, q = 1, 2$, $L_{pq}$ is a tensor having components $L_{pq}^{ij}$, where $i, j = x, y, z$.

character of $\hat{\rho}(\boldsymbol{H})$ can be derived from the appropriate Onsager relation by taking into account that time-reversal transformations switch the direction of the magnetic field. Similarly, one can show that the first nonvanishing term in the magnetic field dependence of the diagonal element is second-order in $H$:

$$\rho(\boldsymbol{H}) = \rho_0 + \beta_1 \boldsymbol{H}^2 + \beta_2 \boldsymbol{H}(\boldsymbol{j} \cdot \boldsymbol{H})/|\boldsymbol{j}| . \qquad (I.7.6)$$

The coefficients $\beta_1$ and $\beta_2$ characterize the *magnetoresistance* – that is, the change in resistivity due to the application of magnetic field. Eq. I.7.6 illustrates that the transverse magnetoresistance (measured with the magnetic field perpendicular to the current) and the longitudinal magnetoresistance (in magnetic field parallel to the current) may be different.

Many more transport coefficients can be defined in a finite temperature gradient and a magnetic field. For example, there is a nonzero coefficient relating the the generation of a temperature gradient in the $y$-direction by a current flow in the $y$-direction (the Ettingshausen effect). For a more general discussion of *thermomagnetic effects* see Callen [10] pp. 305–307, or Ziman [14] pp. 495–501.

In crystalline materials, the transport coefficients are often anisotropic even if $H = 0$. For simplicity, let us consider the the isothermal electric conductivity. Since the electric field and current are vectors, the most general linear relation between them is described by a tensor: $\boldsymbol{j} = \hat{\sigma}\boldsymbol{E}$ or $\boldsymbol{E} = \hat{\rho}\boldsymbol{j}$, where $\hat{\sigma}$ and $\hat{\rho}$ are the conductivity and resistivity tensors, respectively. The number of independent components of the tensors are constrained by the requirement that the physical properties must not change when the crystal is subjected to a symmetry operation. Furthermore, the Onsager relation ensures that the conductivity tensor is symmetric.[3] The $\hat{\sigma}$ tensor is also positive definite, so that the power dissipation, $P = \boldsymbol{j}\boldsymbol{E}$, is always positive.

The *Drude model* is a widely used, phenomenological approach en route towards a true microscopic theory of transport: The particles are characterized by a single effective mass $m$, the interactions are represented by a single relaxation time $\tau$, and the underlying picture is that of a classical ideal gas. In the Drude model of the electronic transport, the conductivity and the Hall coefficients are[4]

$$\sigma = \frac{ne^2\tau}{m} = \frac{1}{3}e^2 g v_{\mathrm{F}}{}^2 \tau \qquad (I.7.7)$$

$$R_{\mathrm{H}} = -\frac{1}{ne} , \qquad (I.7.8)$$

where $n$ is the number density of the electrons, and the negative sign is due to the electrons' negative charge. [In Eq. I.7.7 we also expressed the conductivity

---

[3] As Eq. I.7.4 suggests, this is true only if the magnetic field is zero. Anisotropic materials in a magnetic field are discussed by Landau and Lifshitz [9] Vol. 8, pp. 87–91.

[4] For further details, see Ashcroft and Mermin [1] p. 1.

in terms of the density of states at the Fermi level, $g(E_F)$, and Fermi velocity, $v_F$, using the free-electron values in terms of $n$ and $m$.] These expressions work surprisingly well for metals and (with clever choices of $n$ and the charge) for semiconductors. Similarly, the Drude result for the thermal conductivity of metals,

$$\kappa_e = \frac{3}{2} \frac{k_B}{e}^2 T\sigma, \qquad (\text{I.7.9})$$

is close to reality (only when written in terms of the electrical conductivity, as above). For phonons a similar model yields

$$\kappa_{ph} = \frac{1}{3}cv^2\tau = \frac{1}{3}cv\ell , \qquad (\text{I.7.10})$$

where $c$ is the specific heat, $v$ is a typical phonon velocity, and the *mean free path* $\ell$ is defined as $\ell = v\tau$.

To obtain estimates of the thermoelectric power, it is often useful to think of it as $S = \Pi/T = j_Q/j_eT$. Under the influence of an electric field or a temperature gradient the particles transporting heat and charge start to move with some drift velocity $v_d$. A crude estimate of the heat current is obtained by taking $j_Q = \Delta Q v_d = n(\int cdT)v_d$; for the electric current, $j_e = v_d ne$. These yield $S = (\int cdT)/(Te)$. For free-electrons we then obtain

$$S \approx \frac{\pi^2}{4} \frac{k_B}{e} \frac{k_B T}{E_F} \quad (\text{metals}) , \qquad (\text{I.7.11})$$

again in reasonable agreement with experiment. In a disordered electronic conductor, exhibiting *hopping conductivity*, the electrons carry a constant entropy of the order of $\ln 2$; the electronic specific heat is independent of temperature, and the thermopower

$$S \approx \ln 2 \, k_B/e = 60 \text{ mV/K} \quad (\text{hopping conductors}) \qquad (\text{I.7.12})$$

is also independent of temperature. In semiconductors, the electrons must be excited across the energy gap $E_g$ and the corresponding energy quanta are dissipated in the form of heat. Consequently the heat current is $j_Q = E_g n v_d$ and the thermopower becomes[5]

$$S \approx (k_B/e)(E_g/k_B T) \quad (\text{semiconductors}) . \qquad (\text{I.7.13})$$

The true microscopic theory of transport properties is based on the *Boltzmann equation*. The underlying assumption of the Boltzmann equation is that large portions of the system can still be described by a distribution function similar to the equilibrium distribution function $f(E(k)) \equiv f(k)$, as discussed in Eqs. I.6.21 – I.6.23. But to account for the inhomogeneities created by the external perturbation, the distribution function is made position-dependent. (For an introductory discussion see Ashcroft and Mermin [1] pp. 316–320,

---

[5] These types of arguments are used extensively by Mott [15] and by Ziman [3].

Ziman [3] pp. 211–213, Ibach and Lüth [4] pp. 168–170, or Harrison [5] pp. 253–255. For a more detailed survey, see Ref. [14].) In its most general (and least useful) form the Boltzmann transport equation summarizes the balance between the various ways the distribution function can change:

$$\left.\frac{\partial f}{\partial t}\right|_{\text{field}} + \left.\frac{\partial f}{\partial t}\right|_{\text{diff}} + \left.\frac{\partial f}{\partial t}\right|_{\text{scatt}} = 0 \ . \tag{I.7.14}$$

where the terms correspond to field induced motion, diffusion, and scattering (or collisions) of particles. For electrons the first term is related to the externally applied electric and magnetic fields, the diffusion term is due to the free propagation of the particles in a perfectly periodic crystal field, and the collision term describes the interaction of electrons with lattice imperfections (impurities, dislocations, lattice vibrations) and with other electrons. For phonons the first term is usually zero.

The *linearized Boltzmann equation* is obtained if we assume that $f$ is close to the equilibrium distribution function $f^0$ and the difference, $\delta f = f - f^0$, is small. The first two terms in Eq. I.7.14 can be treated by appropriate expansion of the $f^0$ function, as discussed by several textbooks. The scattering term represents the greatest challenge. To obtain the total change in $f$, one has to consider all processes taking away from and adding to the particle number for state $|k\rangle$. If the transition probability from state $|k\rangle$ to state $|k'\rangle$ is denoted by $W_{kk'}$, then one obtains

$$\left.\frac{\partial f}{\partial t}\right|_{\text{scatt}} = \int \frac{d\mathbf{k}'}{(2\pi)^3} W_{kk'}[f(k) - f(k')] \ . \tag{I.7.15}$$

The principle of *microscopic reversibility*, $W_{kk'} = W_{k'k}$, was used.[6] However, even with that simplification Eq. I.7.15 turns Eq. I.7.14 into an integro-differential equation which is hard to solve. Most often the *relaxation time approximation* is used:

$$\left.\frac{\partial f}{\partial t}\right|_{\text{scatt}} = -\frac{1}{\tau(k)} \, \delta f(k) \ . \tag{I.7.16}$$

Some of the complex wavenumber dependence of $W_{kk'}$ may be condensed into a wavenumber-dependent relaxation time, $\tau(k)$; sometimes this is further simplified to $\tau \equiv \tau(E)$ or to a single relaxation time $\tau$. For example, when the scattering is isotropic and elastic, the relaxation time can be calculated (by comparing Eqs. I.7.16 and I.7.15):

$$\frac{1}{\tau(k)} = \int \frac{d\mathbf{k}'}{(2\pi)^3} W_{kk'} (1 - \cos\Theta) \ , \tag{I.7.17}$$

---

[6] The scattering of a single particle by an impurity is often calculated by using the Golden Rule, $W_{kk'} = 2\pi/\hbar \delta(E(k) - E(k'))|\langle k|U|k'\rangle|^2$, where $U$ is the perturbation potential. As the Dirac $\delta$ function indicates, this scattering is elastic. The condition for microscopic reversibility is also satisfied. For a discussion of the Golden Rule, see Landau and Lifshitz [9] Vol. 3.

where $\Theta$ is the angle between $k$ and $k'$.

To complete the microscopic description of the transport properties, we have to define how currents are calculated from the distribution function:

$$j_n = \int \frac{2dk}{(2\pi)^3} f(k)v(k) \qquad \text{Particle current}$$

$$j_e = e \int \frac{2dk}{(2\pi)^3} f(k)v(k) \qquad \text{Electric current} \qquad \text{(I.7.18)}$$

$$j_E = \int \frac{2dk}{(2\pi)^3} f(k)E(k)v(k) \qquad \text{Energy current}$$

$$j_Q = \int \frac{2dk}{(2\pi)^3} f(k)(E(k) - \mu)v(k) \qquad \text{Heat current.}$$

For the last equation we used $\Delta Q = TdS = E - \mu dN$ from Eq. I.6.1. The integrations are over the first Brillouin zone. Except for the electric current and the difference in the distribution functions, similar equations work for phonons and other quasiparticles as well. Due to the $f^0(k) = f^0(-k)$ symmetry of the Fermi function,[7] we may use $\delta f = f - f^0$ in Eqs. I.7.18. By substituting the solution of the Boltzmann equation into the above currents, the transport coefficients can be evaluated. For example, the conductivity tensor is

$$\sigma_{ij} = e^2 \int \frac{dk}{4\pi^3} \left( \frac{-\partial f^0}{\partial E} \right) \tau v_i(k)v_j(k) . \qquad \text{(I.7.19)}$$

In general, the thermoelectric transport coefficients defined in Eq. I.7.1 are

$$L_{11} = e^2 \int \frac{dk}{4\pi^3} \left( \frac{-\partial f^0}{\partial E} \right) \tau \, v(k) \circ v(k)$$

$$L_{12} = L_{21} = -Te \int \frac{dk}{4\pi^3} \left( \frac{-\partial f^0}{\partial E} \right) \tau \, v(k) \circ v_j(k)(E(k) - \mu)$$

$$L_{22} = T \int \frac{dk}{4\pi^3} \left( \frac{-\partial f^0}{\partial E} \right) \tau \, v(k) \circ v_j(k)(E(k) - \mu)^2 . \qquad \text{(I.7.20)}$$

Here we used the "dyadic product" $(v \circ v)_{ij} \equiv v_i v_j$ to define each component of the L tensors.[8] Note that the Onsager relation (Eq. I.7.2) is automatically satisfied. The integrals can be evaluated in terms of the Bethe–Sommerfeld expansion (Eqs. I.6.30 and I.6.31). Keeping the first nonvanishing terms results in

---

[7] This symmetry relies on the time-reversal symmetry of the Schrödinger equation, which leads to $E(k) = E(-k)$. In most cases the unperturbed system is in thermal equilibrium (described by the equilibrium distribution function of the quasiparticles, $f^0$) and does not carry any current. Notable exceptions are superconductors and superfluids.

[8] For isotropic systems, the L tensors are described by a diagonal matrix with the diagonal components $L_{xx} = L_{yy} = L_{zz}$. In this case the identity $v^2 = v_x^2 + v_y^2 + v_z^2$ allows us to replace $v \circ v$ by $\frac{1}{3}v^2$, and the L tensors may be replaced by scalars.

$$L_{11} = \hat{\sigma} = e^2\tau \int \frac{d\mathbf{k}}{4\pi^3}\delta(E(\mathbf{k})-E_F)\mathbf{v}(\mathbf{k})\circ\mathbf{v}(\mathbf{k})$$

$$L_{12} = -\frac{\pi^2}{3e}(k_BT)^2\{\frac{1}{\tau}\frac{d\tau}{dE}\hat{\sigma}+e^2\tau\int\frac{d\mathbf{k}}{4\pi^3}\delta(E(\mathbf{k})-E_F)\frac{\partial^2 E}{\partial \mathbf{k}\partial \mathbf{k}}\}$$

$$L_{12} = \frac{\pi^2}{3e}(k_BT)^2\hat{\sigma} . \tag{I.7.21}$$

Here we have allowed for an energy-dependent relaxation time. In this approximation (which is well-justified for the degenerate Fermi gas, with $T_F \gg T$) the thermal conductivity in Eq. I.7.3 reduces to $\kappa = L_{22}$, and it satisfies the *Wiedemann–Franz law*: $\kappa = \alpha T\sigma$, where the constant of proportionality is $\alpha = (\pi^2/3)(k_B/e)^2$. The thermoelectric power is $S = -\frac{\pi^2}{3}(k_B/e)k_BT/E^0$, where $E^0$ depends on the band structure and can be evaluated from Eq. I.7.21. $E^0 \approx E_F$ is a good estimate. For free electrons these results justify the Drude results, Eqs. I.7.7–I.7.9, and the simple estimate for the thermopower, Eq. I.7.11. However, the agreement is in part due to the fact that we did not solve the "real" Boltzmann equation after all; instead we used the relaxation time approximation.

The electrical resistivity of metals has an important contribution due to electron–phonon scattering. A simple estimate yields a relaxation rate proportional to the number of phonons $N_{ph}$. At high temperatures $N_{ph}$ varies linearly with temperature (see Problem 6.2), leading to the observed linear temperature dependence of the resistance. A more sophisticated calculation takes into account the directional dependence (Eq. I.7.17) and the inelastic nature of the scattering. The result is the Bloch–Grüneisen formula:

$$\rho \sim \frac{T}{\Theta_D}\int_0^{T/\Theta_D}\frac{z^5 dz}{(e^z-1)(1-e^{-z})} , \tag{I.7.22}$$

where $\Theta_D$ is the Debye temperature. This function is tabulated in Landolt–Börnstein [21] Vol. 15, p. 287.

If the relaxation time $\tau$ is independent of the electron wavenumber the solution of the Boltzmann transport equation gives an electrical conductivity tensor

$$\sigma_{i,j} = \frac{1}{4\pi^3}\frac{e^2\tau}{\hbar}\int\frac{v_i v_j dS_F}{|v|} . \tag{I.7.23}$$

$v_i$ is the $i$th component of the Fermi velocity, and $|v|$ is the absolute value of $v$. The integration is over the Fermi surface.

The Hall effect and the magnetoresistance can be treated similarly. The solution of the Boltzmann equation is searched for in terms of a power series of the operator $(e\tau/\hbar c)[\mathbf{v}\times\mathbf{H}]\,\partial/\partial\mathbf{k}$. Here we reproduce the result for a two-dimensional electronic system. It is assumed that the electronic energy has no dispersion in the direction of the applied magnetic field, which points in the $z$-direction.[9]

$$\sigma_{xy} = \frac{2e^3}{\hbar} \int \frac{d\mathbf{k}}{4\pi^3} \left( \frac{-\partial f^0}{\partial E} \right) v_y \tau(\mathbf{k}) \left\{ v_y \frac{\partial}{\partial k_x} - v_x \frac{\partial}{\partial k_y} \right\} v_x \tau(\mathbf{k}) . \qquad \text{(I.7.24)}$$

Note that the momentum-dependence of the relaxation rate is retained here.

## 7.1 Problem: Temperature Dependent Resistance

Resistors made of pure Pt metal are often used to measure temperature down to about 20 K (below that temperature the sensitivity dramatically decreases). In Table I.7.1 we reproduced the calibration points for a standard, "100 $\Omega$" Pt resistor. Show that the Bloch–Grüneisen formula gives a satisfactory fit to these data for temperatures below 350 K. (The Debye temperature of Pt is 230 K. The Bloch–Grüneisen function, Eq. I.7.22, is tabulated in Landolt–Börnstein [21] Vol. 15, p. 287. It can also be calculated by numerical integration.)

Table I.7.1. Calibration data for a typical "100 $\Omega$" Pt resistor.

| T(K) | R($\Omega$) | T(K) | R($\Omega$) |
|------|------|------|------|
| 14.0 | 1.797 | 100.0 | 29.987 |
| 20.0 | 2.147 | 150.0 | 50.815 |
| 30.0 | 3.508 | 200.0 | 71.073 |
| 40.0 | 5.938 | 300.0 | 110.45 |
| 50.0 | 9.228 | 400.0 | 148.62 |
| 70.0 | 17.128 | 1000.0 | 353.402 |

## 7.2 Problem: Conductivity Tensor

Prove that for a tetragonal crystal the conductivity is isotropic in the plane perpendicular to the $c$ axis. (Note how powerful a statement this is. For example, in an electrical conductivity measurement one obtains the same value if the current flows along the CuO bonds of Figure I.1.6 or if it flows in a direction 45° to them.)

---

[9] The more general result, and details of the derivation, can be found in Ziman [14] pp. 501–504. The two-dimensional formula is quoted by Ong [22], who also provides a transparent geometrical representation of the magnetotransport in two-dimensional systems.

## 7.3 Problem: Montgomery Method

The resistivity of anisotropic samples is studied by using the Montgomery method [23]. Assume that a thin, crystalline specimen is cut to square shape, so that the edges coincide with the principal axis of the resistivity tensor $\rho$. There are four contacts at the corners of the sample (Figure I.7.1). Dimensions $d_1$, $d_2$, and (perpendicular to the specimen) $d_3$ are $d_1 = d_2 = 1$ mm and $d_3 = 0.01$ mm. In this geometry two elements of the resistivity tensor, $\rho_1$ and $\rho_2$, can be determined by sending current and measuring voltage between various combinations of the contacts.

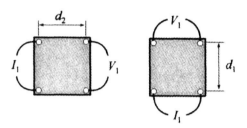

**Fig. I.7.1.** Sketch of the geometry used in the Montgomery method of measuring resistivity.

The resistances $R_1$ and $R_2$ are defined as the ratio of measured voltages and currents in configurations indicated in the figure. $R_1 = 10\ \Omega$ and $R_2 = 20\ \Omega$ are obtained from the data. Use Montgomery's work to determine $\rho_1$ and $\rho_2$. How would you change the sample geometry if the resistivity anisotropy is expected to be in the range of $\rho_2/\rho_1 \sim 20$?

## 7.4 Problem: Anisotropic Layer

Samples are cut from a single-crystal specimen in four different directions and are labeled 1, 2, 3, and 4 (see Figure I.7.2). Each cut has the same length and cross section; the width and thickness of each is much less than its length. The resistance of cuts 1 and 3 are measured to be equal, $R_1 = R_3 = R$; the resistance of cut 2 is $R_2 = R/2$. What is the resistance of cut 4?

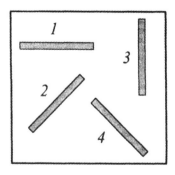

**Fig. I.7.2.** The four samples, labeled 1, 2, 3, and 4, cut from an anisotropic material. There is a 45° increment between the subsequent cuts.

## 7.5 Problem: Two-Charge-Carrier Drude Model

Consider a system of two types of charge carriers in the Drude model. The two carriers have the same density $(n)$ and opposite charge ($e$ and $-e$), and their masses and relaxation rates are $m_1, m_2$ and $\tau_1, \tau_2$, respectively. (You may want to use the mobility, $\mu = \tau/m$, instead of $\tau$ and $m$.)

(a) Calculate the magnetoresistance, $\Delta\rho = \rho(H) - \rho(H = 0)$, where $H$ is the magnetic field.

(b) Calculate the Hall coefficient.

(c) In an undoped semiconductor, $n = n_0 e^{-\Delta/k_B T}$ describes the temperature dependence of the carrier concentration. What will the temperature dependence of the magnetoresistance and the Hall coefficient be?

## 7.6 Problem: Thermal Conductivity

We perform a thermal conductivity measurement (i.e., we establish a $\nabla T$ temperature gradient, and we measure the $j_q$ heat flow across the sample), but instead of the standard zero electrical current condition, we enforce zero electric field across the sample. We calculate $\kappa' = -\frac{j_q}{\nabla T}$ from the measured heat flow.

(a) Calculate the difference $\Delta\kappa = \kappa' - \kappa$ in terms of the usual thermal conductivity $(\kappa)$, conductivity $(\sigma)$, and thermopower $(S)$.

(b) How big is $\Delta\kappa/\kappa$ at room temperature for a typical metal?

## 7.7 Problem: Residual Resistivity

According to the *Matthiessen rule* the residual resistivity of dilute alloys is proportional to the concentration of impurities:

$$\rho_0 = \alpha c_i . \tag{I.7.25}$$

In Figure I.7.3 the coefficient $\alpha$ is shown for alloys of copper with various other materials.[10] Estimate the scattering rate of the electrons, calculate the residual resistivity, and compare the results to the experiment.

---

[10] The data were measured by F. Pawlek and K. Riecher, *Z. Metallkunde*, **47**, 347 (1956) and discussed in detail by J. Friedel, *Can. J. Phys.*, **34** 1190 (1956).

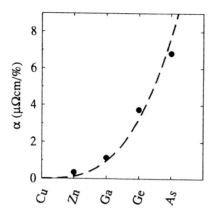

**Fig. I.7.3.** Residual resistivity of various copper alloys after Pawlek and Riecher. On the horizontal axis are the various dopants in order of increasing atomic number. On the vertical axis, the coefficient $\alpha$ in Eq. I.7.25 is plotted. The dashed line represents a fit to the data, assuming that $\alpha \sim Z^2$, where $Z$ is the deviation of the atomic number away from copper.

## 7.8 Problem: Electric and Heat Transport

Use your knowledge of metals and semiconductors to do the following:

(a) Sketch the temperature dependence of the resistivity for a metal and for an intrinsic semiconductor. Indicate the characteristic temperature dependencies and the order of magnitude numerical values, and identify the physical origins of the various terms.

(b) Now do the same for the thermal conductivity of metals and insulators. Scale the temperature axis only.

## 7.9 Problem: Conductivity of Tight-Binding Band

Consider a two-dimensional metal with a square lattice of lattice spacing $a$. The conduction band is described by the tight-binding approximation,

$$ E = E_0 + E_1 \left(2 - \cos k_x a - \cos k_y a\right) , \qquad (\text{I.7.26}) $$

and the relaxation time $\tau$ is independent of the electron momentum or energy. The band is half-filled.

(a) Using the solution of the Boltzmann equation, calculate the conductivity tensor.

(b) Compare the result of (a) to the conductivity from the Drude model. (Use the same electron density and relaxation time. What is the relevant effective mass?)

## 7.10 Problem: Hall Effect in Two-Dimensional Metals

The conduction electron band in a two-dimensional metal is described by

$$E = E_0 + E_1(2 - \cos k_x a - \cos k_y a) , \qquad (I.7.27)$$

where $a$ is the lattice spacing. The relaxation time $\tau$ is independent of the electron momentum or energy. Determine the Hall coefficient for a nearly empty, a half-filled, and a nearly full band and discuss the result. (*Note:* Results from a paper by N.P. Ong [22] may be used for this problem without proof.)

## 7.11 Problem: Free-Electron Results from the Boltzmann Equations

Show that for "free" electrons the conductivity (Eq. I.7.23) reduces to

$$\sigma = \frac{ne^2\tau}{m} , \qquad (I.7.28)$$

where $n$ is the density of electrons and $m$ is the electron mass.

## 7.12 Problem: $p$–$n$ Junctions

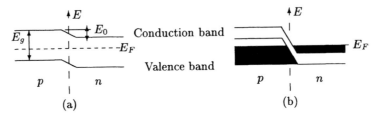

(a)                                                              (b)

**Fig. I.7.4.** (a) Energy levels for an unbiased $p$–$n$ junction. The contact potential (the energy level shift at the $p$–$n$ interface) is given by $E_0$; $E_F$ is the Fermi energy. (b) Energy levels of an unbiased heavily doped junction where tunneling can now occur.

(a) The junction between a $p$-type and an $n$-type semiconductor is regularly used as an electronic device; the energy levels for such a device are drawn in Figure I.7.4(a).
   The zero bias diffusion currents for electrons and holes are defined as

$$I_{pn}^e, \ I_{pn}^h, \ I_{np}^e, \ I_{np}^h , \qquad (I.7.29)$$

where the superscript denotes electron ($e$) or hole ($h$) current and the subscript denotes the direction of the current, $p$ to $n$ ($pn$) or $n$ to $p$ ($np$). The net current must be zero when the bias is zero, so we know that

$$I^e_{pn} = -I^e_{np} \quad \text{and} \quad I^h_{pn} = -I^h_{np} \ . \tag{I.7.30}$$

If we now bias the junction with voltage $V$ (by putting the positive voltage lead on the $p$ side and the negative lead on the $n$ side) the majority carrier currents will become

$$I^e_{np}(V) = I^e_{np}e^{eV/K_BT} \quad \text{and} \quad I^h_{pn}(V) = I^h_{pn}e^{eV/K_BT} \ , \tag{I.7.31}$$

while the minority carrier currents remain unchanged. We are assuming that $eV < E_0$. Plot the total current for this junction as a function of $V$ (for both positive and negative voltages). What is this device?

(b) When semiconductors are very heavily doped, the depletion region at the junction can become quite narrow. In such a situation, as shown in Figure I.7.4(b), tunneling will occur when possible.

Without any calculations, plot the $I$–$V$ curve for such a device.

# 8 Optical Properties

The interactions between electromagnetic radiation and condensed matter are extensively used to study a material's properties – after all, even our own eyes analyze reflected and scattered electromagnetic waves. In condensed matter physics the terminology of optical response is extended to a wide range of frequencies, from microwave to ultraviolet (UV) light. At extremely low frequencies, the response is determined by the steady-state dynamic equilibrium, as discussed in great detail in Chapter 7; at the high frequency end, when the wavelength becomes comparable to the interatomic spacing, the methods and terminology of X-ray diffraction (see Chapter 1) are more appropriate.

Throughout our discussion we will assume that the applied electric and magnetic fields are small and the response of the system is linear. It is then useful to view an arbitrary time-dependent field as a superposition of harmonic waves, with amplitudes determined by a Fourier transformation. The response of the system will then be the superposition of the responses to each of the harmonic components. The electric field ($E$) is written

$$E(t) = E(k, \omega)e^{i(kr - \omega t)} , \tag{I.8.1}$$

where $E(k, \omega)$ is the (complex) amplitude of the oscillation. Similar formulae are used to describe the time dependence of the electric polarization ($P$), the current density ($j$), the charge density ($\rho$), and other quantities.

The optical response of a material is characterized by the *optical conductivity* $\sigma(k, \omega)$, *dielectric susceptibility* $\chi(k, \omega)$, or the *dielectric function* $\epsilon(k, \omega)$. These quantities are defined as the coefficients in the equations

$$\begin{align}
j(k, \omega) &= \sigma E(k, \omega) \tag{I.8.2}\\
P(k, \omega) &= \chi E(k, \omega) \tag{I.8.3}\\
D(k, \omega) &= \epsilon E(k, \omega) , \tag{I.8.4}
\end{align}$$

and they are, in general, tensors. By definition, the electric displacement vector is $D = E + 4\pi P$ and therefore $\epsilon = 1 + 4\pi\chi$. These equations imply another assumption, namely that the current (polarization) at any point in the material depends on the electric field at that point only. This assumption is valid as long as the wavelength of the radiation is much longer than the length scale characterizing the electronic response. There are two notable cases when the assumption is *not* valid: in pure metals at low temperatures, where the mean free path of the electrons may become longer than the decay rate of the electromagnetic field, in which case the *anomalous skin effect* occurs; and in *type I superconductors*, where the correlation length of the Cooper pairs is larger than the penetration depth of the magnetic field. In the present discussions we will confine ourselves to local responses.

Inserting Eqs. I.8.2, I.8.4, and I.8.1 into Maxwell's equations results in

$$k(kE) - k^2 E \; = \; -\frac{1}{c^2}\left(4\pi i\omega\sigma + \epsilon\omega^2\right)E$$

$$= \; -\frac{\omega^2}{c^2}\left(\frac{4\pi i\sigma}{\omega} + \epsilon\right)E \; . \qquad (I.8.5)$$

Note that the conductivity and the dielectric function enter in these equations only as a combination: $\epsilon + \frac{4\pi i\sigma}{\omega}$. The dielectric function is used to describe the electronic response of "bound" charges (like electrons on an localized atomic orbit, or positive and negative ions in a solid); the conductivity is more convenient for "free" carriers (like conduction electrons in a metal or semiconductor, or moving ions in a superionic conductor). However, from the perspective of optical properties, this distinction becomes irrelevant: In an oscillating electric field the free charges will perform oscillatory motion just like the bound charges. Instead of using $\epsilon$ and $\sigma$, we can characterize the response of the material by the generalized dielectric function, defined as

$$\epsilon^* = \epsilon + i\frac{4\pi\sigma}{\omega} \; . \qquad (I.8.6)$$

Alternatively, we can use the generalized conductivity ($\sigma^*$), defined by the equation

$$\sigma^* = i\omega\frac{1 - \epsilon^*}{4\pi} \; . \qquad (I.8.7)$$

In $\sigma^*$, the 1 accounts for the fact that if no material is around, the dielectric constant is unity and the conductivity must be zero.[1] The equations above suggest that $\epsilon^*$ and $\sigma^*$ are complex functions of $\omega$. The real and imaginary parts of these functions will be denoted by subscripts 1 and 2, respectively: $\epsilon_1 \equiv \mathrm{Re}(\epsilon)$ and $\epsilon_2 \equiv \mathrm{Im}(\epsilon)$.

At this point we want to consider the transverse ($k \perp E$) and longitudinal ($k \parallel E$) waves. For transverse waves, $kE = 0$, Eq. I.8.5 becomes

$$\left[k^2 - \frac{\omega^2}{c^2}\left(\epsilon + \frac{i4\pi\sigma}{\omega}\right)\right]E = 0 \; . \qquad (I.8.8)$$

This equation has a nonzero solution for the electric field if the coefficient is zero:

$$k = \frac{\omega}{c}\sqrt{\epsilon + \frac{i4\pi\sigma}{\omega}} = \frac{\omega}{c}\sqrt{\epsilon^*} \; . \qquad (I.8.9)$$

---

[1] For further discussion see Ashcroft and Mermin [1] p.776.

The factor $n \equiv n + i\kappa = \sqrt{\epsilon^*}$ is called the complex *index of refraction*. If $\epsilon^*$ is a positive real number, than the solution corresponds to waves propagating in the medium (with a reduced velocity, $c' = c/n$). If $\epsilon^*$ is complex, or real but less then zero, the amplitude of the wave varies and energy cannot propagate through the medium.

Alternatively, for longitudinal waves, $\mathbf{k}(\mathbf{k}\mathbf{E}) = k^2\mathbf{E}$ and we require (see Eqs. I.8.5)

$$\epsilon + \frac{4\pi i \sigma}{\omega} = 0 , \tag{I.8.10}$$

or simply $\epsilon^* = 0$. In an insulator the longitudinal polarization waves, coupled to electromagnetic radiation, are called *polaritons*;[2] similar longitudinal waves in a metal are *plasmons*. In contrast to Eq. I.8.9, the wavenumber dependence is not explicit in Eq. I.8.10. However, in general, $\epsilon = \epsilon(\mathbf{k}, \omega)$ and $\sigma = \sigma(\mathbf{k}, \omega)$ and the two equations have to be solved accordingly.

Usually the optical conductivity or the dielectric function cannot be measured directly. In the most common experimental arrangement, the material is exposed to the electromagnetic radiation, and the *reflectivity* is measured. This can be experimentally achieved either by using freely propagating waves (for infrared and optical radiations) or by using waveguides (for microwave and millimeter waves). Other methods involve measurements of the *transmission coefficient* or the *absorption coefficient* of the sample. In microwave studies the data are often evaluated in terms of the *surface impedance*. The relationship between these quantities and the dielectric function is found when we solve the problem of matching the electromagnetic field at the interface between vacuum (or air) and the sample. A few representative results are summarized here:

– Reflection coefficient for infinite medium and radiation having perpendicular incidence:

$$R \equiv \frac{\text{reflected power}}{\text{incident power}} = \left|\frac{\sqrt{\epsilon} - 1}{\sqrt{\epsilon} + 1}\right|^2 = \frac{(1-n)^2 + \kappa^2}{(1+n)^2 + \kappa^2} . \tag{I.8.11}$$

– Transmission coefficient[3] of a thin metallic or superconducting film of thickness $d$:

$$T \equiv \frac{\text{transmitted power}}{\text{incident power}} \approx \frac{1}{(1 + dZ_0\sigma_1/2)^2 + (dZ_0\sigma_2/2)^2} , \tag{I.8.12}$$

where $\sigma_1$ and $\sigma_2$ are the real and imaginary parts of the conductivity, respectively, and $Z_0 = 4\pi/c$ in CGS units or $Z_0 = 377\ \Omega$ in MKSA units.

---

[2] Experimental and theoretical aspects of polaritons in semiconductors are discussed in detail by Yu and Cardona [6] pp. 282–288.
[3] This formula was first derived by Tinkham [20]; it is only valid for metallic samples.

– Surface impedance of a metal:

$$Z = \frac{4\pi}{c\sqrt{\epsilon^*}} \approx \frac{1-i}{c}\sqrt{\frac{2\pi\omega}{\sigma}} \,. \tag{I.8.13}$$

– According to Eq. I.8.9, when $\epsilon^*$ is negative or complex the amplitude of the electric field oscillations varies with the distance from the surface ($x$) like $e^{\pm x/\lambda}$, where

$$\lambda = \frac{|\delta|^2}{\mathrm{Im}\delta} \qquad \text{and} \qquad \delta = \frac{c}{\sqrt{\epsilon^*}\omega} \,. \tag{I.8.14}$$

In superconductors the quantity $\lambda$ is called the *penetration depth*; in normal metals it is called the *skin depth*.

– Equation I.8.9 can be also used to calculate the *absorption coefficient*: the relative power absorption per unit length. This quantity is meaningful if $n \gg \kappa$ and the wave is decaying over a distance corresponding to many wavelengths. The result is

$$\eta \equiv \frac{\text{absorbed power}}{\text{total power} \times \text{unit length}} = \frac{2\kappa\omega}{c} \,. \tag{I.8.15}$$

The complex functions characterizing the optical properties must be analytical functions of $\omega$ for the upper half of the complex plane above the real axis. This follows from the principle of causality, as discussed by Ziman [3], p. 259, Ibach and Lüth [4], p. 241, and other textbooks. Therefore, the real and imaginary parts are related by the *Kramers–Kronig* transformations. For the dielectric function, they are

$$\begin{aligned}
\epsilon_1^* &= \frac{2}{\pi}P\int_0^\infty \frac{\omega'\epsilon_2^*(\omega')}{\omega'^2 - \omega^2}\,d\omega' + 1 \\
\epsilon_2^* &= -\frac{2\omega}{\pi}P\int_0^\infty \frac{\epsilon_1^*(\omega')}{\omega'^2 - \omega^2}\,d\omega' \,,
\end{aligned} \tag{I.8.16}$$

where the $P$ stands for the "principal" part of the integral which amounts to an appropriate handling of divergences during the integration. Similar equations hold for real and imaginary parts of $\sigma(\omega)$, $\epsilon(\omega)$, $\sigma^*(\omega)$, $n(\omega)$, $Z(\omega)$, and so on.

Another important restriction is placed on the dielectric function by the *oscillator sum rule*:

$$\frac{1}{4\pi}\int_0^\infty \omega\epsilon_2^*\,d\omega = \int_0^\infty \sigma_1^*\,d\omega = \frac{\pi}{2}\frac{Ne^2}{m} \,, \tag{I.8.17}$$

where $n$ is the density of the electrons in the system and $m$ is the mass of the electrons.[4]

---

[4] See, for example, Landau and Lifshitz [9] Vol. 8, p. 282.

The calculation of the dielectric function or the conductivity is based on a microscopic theory describing the motion of the electrons. For the calculation of the optical response in the simplest way, the charge carriers are treated as classical particles. To eliminate the divergence of the response, a phenomenological damping force must be included. This model (the *Drude model*) works quite well for metals and semiconductors, as long as the energy quanta of the electromagnetic radiation are smaller than the interband transition energy of the electrons within the material. For insulators a restoring force is also added, and a resonance frequency is often set so that it matches the resonance transition between two relevant energy bands.

For an electric charge $e$, mass $m$, damping $\gamma$, and restoring force constant $= m\omega_0^2$, the solution of the classical equation of motion yields $x = (Ee/m)(\omega_0^2 - \omega^2 - i\omega/\Gamma)^{-1}$. The electric dipole moment of each electron is $p = ex$ and the polarization of the medium is $P = 4\pi np$, where $n$ is the density of the electrons. The dielectric function is then

$$\epsilon^* = 1 + \frac{4\pi ne^2}{m} \frac{1}{\omega_0^2 - \omega^2 - i\omega\Gamma} , \qquad (I.8.18)$$

where the symbol $\Gamma \equiv \gamma/m$ was used. For $\Gamma \ll \omega_0$ the dielectric function obtained here describes sharp resonances, with a divergent response in the neighborhood of $\omega_0$.

Equation I.8.18 can be generalized for dipole transitions between more than two energy levels (see, for example, Ziman [3] p. 265),

$$\epsilon^* = 1 + \frac{4\pi ne^2}{m} \sum_j \frac{f_j}{\omega_j^2 - \omega^2 - i\omega\Gamma_j} , \qquad (I.8.19)$$

where $\hbar\omega_j$ is the energy difference between the bands and $f_j$ is the *oscillator strength* of the $j$th transition. The oscillator strengths can be calculated in first-order perturbation theory by using the initial- and final-state wavefunctions and the electric potential of an approximately homogeneous electric field. Note that the oscillator strengths obey the rule $\sum_j f_j = 1$ so that the oscillator sum rule (Eq. I.8.17) is satisfied. Equation I.8.19 works well for insulators, where the transitions are between narrow energy levels.

When transitions between two bands are excited (see Ziman [3] p. 273 or Ibach and Lüth [4] p. 266) the sharp transition energies in Eq. I.8.19 are replaced by a probability distribution of energies, described by the *joint density of states*:

$$g_{1,2} = Z \int \frac{dS}{8\pi^3} \frac{1}{|\partial(E_1(k) - E_2(k))/\partial k|} , \qquad (I.8.20)$$

where $E_1(k)$ and $E_2(k)$ represent the two energy bands. Typically the bandwidth is larger than $\hbar\Gamma$ and the $\Gamma \to 0$ limit can be taken. The real part of the dielectric function is, by straightforward generalization of Eq. I.8.19,

$$\epsilon_1^* \approx 1 + \frac{4\pi n e^2}{m} \int \frac{f(\omega')g_{1,2}(\omega')}{\omega'^2 - \omega^2} . \tag{I.8.21}$$

Via a Kramers–Kronig transformation, the conductivity is obtained:

$$
\begin{aligned}
\sigma_1^* &\approx \frac{\pi}{2} \frac{4\pi n e^2}{m} \int f(\omega')g_{1,2}(\omega')\delta(\omega - \omega') \, d\omega' \\
&= \frac{2\pi^2 n e^2}{m} f(\omega)g_{1,2}(\omega) .
\end{aligned}
\tag{I.8.22}
$$

Note that Eqs. I.8.21 and I.8.22 are valid only if the wavenumber of the electron does not change during the scattering processes (*direct transitions*). During *indirect* or *phonon-assisted* transitions some electron momentum is transferred to phonons. The smallest separation between two energy bands is called the *direct energy gap* if the electron momentum is not changed, and it is referred to as the *indirect gap* if the electron momentum can vary. The sharp onset of the optical conductivity at the *absorption band edge*, observed in most semiconductors, is due to the van Hove singularity in $g_{1,2}$ at the direct gap energy.

For metals we obtain the conductivity from Eq. I.8.18 by using Eq. I.8.7 and inserting $\omega_0 = 0$,

$$\sigma = \frac{n e^2}{m} \frac{1}{\Gamma - i\omega} = \frac{\sigma_0}{1 - i\omega\tau} . \tag{I.8.23}$$

The *dc* conductivity from Eq. I.7.7, $\sigma_0 = n e^2 \tau / m$, was introduced, with a relaxation time defined as $\tau = 1/\Gamma$. (Identical results are obtained if the conductivity is calculated from $j = \sigma E$, where $j = nev = nedx/dt$.) In this model the solution of Eq. I.8.10 is $\omega \approx \sqrt{4\pi n e^2/m}$. Recalling that the corresponding longitudinal waves are plasmons, the quantity

$$\omega_p = \sqrt{4\pi n e^2/m} \tag{I.8.24}$$

is called the *plasma frequency*.[5] For alkali metals, typical values of the plasma frequency are in the range of $\omega_p = 10^{15}$ sec$^{-1}$; relaxation rates are in the range of $1/\tau = \Gamma = 10^{13}$ sec$^{-1}$.

Equation I.8.23 describes the contribution of conduction electrons to the optical response of semiconductors and metals, assuming that the effective mass approximation works. The solution to the Boltzmann equation[6] treats the band structure better:

$$\sigma_{ij} = \frac{e^2}{2\pi^2} \int \frac{\tau v_i v_j}{1 - i\tau(\omega - \boldsymbol{K} \cdot \boldsymbol{v}_k)} \frac{dS_F}{\hbar |\boldsymbol{v}_k|} . \tag{I.8.25}$$

---

[5] Sometimes the oscillator sum rule, Eq. I.8.17, is written in the form of $\int_0^\infty \sigma_1 \, d\omega = \omega_p^2/8$.

[6] See, for example, Eq. 8.89 in Ziman [3].

The term $\boldsymbol{K} \cdot \boldsymbol{v}$ in the denominator can often be neglected, since the speed of the electromagnetic radiation is larger than the typical electron velocities.

The theoretical understanding and the experimental study of microwave and optical conductivity proved to be particularly important for electronic systems where the interaction between the electrons leads to a new electronic phase at low temperatures. Superconductivity is the most intensely studied and best-known example for an electronic condensate, but charge density waves (CDWs) and spin density waves (SDWs) have been also investigated in great detail. The common feature in these systems is that the interactions lead to a gap in the electronic excitation spectrum, resulting in a strong suppression of the optical conductivity,[7] described approximately by Eq. I.8.22. As long as the oscillator sum rule is valid, the reduction of the optical conductivity at certain frequencies must be accompanied by an increased conductivity at other frequencies. For superconductors, this increase appears as a Dirac $\delta$ function at $\omega = 0$; for certain materials with charge or spin density waves there is a broad resonance structure at finite frequencies. This *collective mode* resonance is discussed by Grüner [16] pp. 164–180.

The optical processes discussed to this point involve the absorption or the emission of light quanta. The corresponding electronic transitions or vibrational modes are called *optically active*[8] or *dipole active*. Classical vibrational motions are optically active if the application of a homogeneous electric field leads to electronic displacements having nonzero electric dipole moments; in a quantum treatment, a nonzero matrix element of the $V = Ex$ potential between the final- and initial-states is required.

There are various vibrational modes or electronic transitions where, due to symmetries, the dipole moment (or the $f_j$ oscillator strength) is exactly zero. These are called *forbidden modes/transitions*. In fact, for solids of complex unit cells, like high-temperature superconductors or materials made with $C_{60}$ molecules, most of the optical modes in the phonon spectrum are forbidden. Group theory is used extensively to sort out the forbidden and active modes [19].

Another aspect to optical properties is related to the inelastic scattering of light: *Raman* and *Brillouin spectroscopy* (see Ibach and Lüth [4] pp. 62–65). Even if an electronic transition or vibrational mode has no dipole activity, it may still be detected by inelastic light scattering. Again, the symmetry group of the crystal can be analyzed to decide which modes are Raman active, and for complex crystals many of the modes will not have Raman activity. Of course, the symmetry considerations relevant for optical spectroscopy are not valid for other types of scattering processes. Neutron or electron scattering can detect "forbidden" optical phonon modes without many restrictions.

---

[7] In the context of superconductivity the oscillator strengths are called *coherence factors*. The calculation of the coherence factors was a great triumph of the BCS theory.

[8] The term "optical activity" is also used in another context to describe the rotation of the polarization of the light.

## 8.1 Problem: Fourier Transform Infrared Spectroscopy

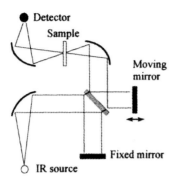

Detector

Sample

Moving
mirror

Fixed mirror

IR source

**Fig. I.8.1.** Schematic sketch of the essential features of a Fourier transform infrared (FTIR) spectrometer.

The Michelson interferometer is the basis of the most widely used spectrometers for infrared (IR) spectroscopy. The spectrometer consists of a "white" light source, a beam splitter sending the light to a fixed mirror and a moving mirror, a sample to be measured, and the detector (see Figure I.8.1). A spectrum is obtained by recording the intensity of the radiation at the detector, as a function of the position of the moving mirror. Usually, two spectra are recorded: one with the sample placed in the path of the light, and another one without the sample. The two spectra are ratioed point by point, and the result is Fourier-transformed to obtain the frequency-dependent transmission of the sample. The reflectivity of the sample can be measured similarly.

(a) Assume that the detector is pointlike and its sensitivity does not depend on the frequency of incident radiation, the source provides uniform intensity at all frequencies, and the displacement $x$ of the moving mirror is measured from the point where the optical path length in the two arms is equal. Show that at finite $\omega$ the quantity

$$T(\omega) = \int_0^\infty I(x) \cos(\alpha\omega x) \, dx \qquad (I.8.26)$$

is indeed proportional to the frequency-dependent transmission of the sample, and determine the constant $\alpha$. (*Note:* In optical spectroscopy the units of wavenumbers, $1/\lambda$, are often used instead of frequency, $\omega = 2\pi c/\lambda$.)

(b) What determines the frequency resolution of the instrument?

(c) The detector signal is amplified by a a device of upper cutoff frequency $f_0 = 100$ kHz. The mirror moves with a velocity $v = 2.5$ cm/sec. What is the high frequency cutoff of the measured spectrum?

## 8.2 Problem: Optical Mode of KBr

Alkali halide crystals must have a strong, optically active resonance due to the oscillation of the two oppositely charged ions in the unit cell. Nevertheless, these compounds are extensively used as window materials in optical spectroscopy for frequencies above 400 cm$^{-1}$ (in units of $1/\lambda = \omega/(2\pi c)$). The bulk modulus of KBr is $B = 1.48 \times 10^{11}$ dyn/cm$^2$, and the lattice spacing is $a = 6.59$ Å. Estimate the frequency of the optical mode at $k = 0$.

## 8.3 Problem: Direct-Gap Semiconductor

Consider a semiconductor with a direct gap of $2\Delta = \hbar\omega_c$ around $k = 0$. The valence band is described by a "parabolic" $E(k)$ with a hole mass of $m_h$; the conduction band is described by an electron mass of $m_e$. Assume that in the neighborhood of $k = 0$, the electronic transition matrix element (for transitions from the valence band to the conduction band) is independent of $k$, and therefore the oscillator strength is constant, $f(k) = f$.

Calculate the conductivity $\sigma_1(\omega)$ around $\omega_c$. What happens in a one-dimensional system?

## 8.4 Problem: Inversion Symmetry

Silicon has acoustic and optical modes in its phonon spectrum. However, the optical modes are not "IR-active" in the sense that IR radiation does not excite these phonons.[9]

(a) Show that the diamond lattice has inversion symmetry. Describe the position of the inversion symmetry points in the crystal.

(b) A vibrational mode is infrared active, if the atomic displacements create an electric dipole moment. Why is the optical mode of silicon not IR active?

## 8.5 Problem: Frequency-Dependent Conductivity

Consider a two-dimensional metal with a square lattice and a half-filled tight-binding band, $E = E_0 + E_1 (\cos k_x a + \cos k_y a)$. Assume that the mean free path, $l$, is independent of the wavenumber of the electrons.

(a) Reduce the solution to the Boltzmann equation (Eq. I.8.25) to a two-dimensional electron system. Express the real and imaginary parts of the

---

[9] These modes are easily observed in Raman or neutron spectroscopy.

frequency-dependent conductivity ($\sigma_1(\omega)$ and $\sigma_2(\omega)$, respectively) in terms of $\omega, l, E_1$, and the lattice spacing $a$.

(b) Calculate the area under the $\sigma_1(\omega)$ curve. Compare the result to the free-electron case. Discuss (qualitatively) why the oscillator sum rule is violated for this single-band model and how it can be restored.

(c) Show that the real and imaginary parts of the conductivity satisfy the Kramers–Kronig relation.

## 8.6 Problem: Frequency-Dependent Response of a Superconductor

Assume that the real part of the conductivity of a superconductor is described by $\sigma_1(\omega) = A\delta(\omega)$. In a nearly free-electron model, how is $A$ related to the electron density, mass and charge? Calculate the penetration depth, $\lambda(\omega)$, and the reflectivity coefficient, $R(\omega)$, for $\omega > 0$. (*Hint:* A way to approximate the Dirac-delta function is to use a Drude conductivity with with a relaxation time $\tau \to \infty$.)

## 8.7 Problem: Transmission of a Thin Superconductor

A thin layer of metal transmits electromagnetic radiation according to the approximation

$$t = \frac{1}{(1 + \frac{dz\sigma_1}{2})^2 + (\frac{dz\sigma_2}{2})^2} ,$$  (I.8.27)

where $t$ is the transmitted fraction of the energy, $\sigma_1$ and $\sigma_2$ are the real and the imaginary part of the conductivity, $d$ is the thickness of the metal, and $z = 377\ \Omega$ [MKSA units] or $4\pi/c$ [CGS units].

(a) Assume that a superconductor is described by $\sigma_1(\omega) = \Omega_p^2 \delta(\omega)$; the Cooper pair condensate results in a delta function at zero frequency. Calculate the transmission coefficient $t$ for low, but nonzero frequencies, and show that the frequency dependence of the transmission can be used to evaluate $\Omega_p$.

(b) Based on the Drude model (with $1/\tau = 0$), estimate the (super)conduction electron density $n$, if $t = 0.02$ for $\lambda = 0.01$-cm-wavelength IR radiation. The thickness of the film is $d = 1000$ Å.

## 8.8 Problem: Bloch Oscillations

Electrons are moving in a semiconductor superlattice, represented by a weak periodic potential of lattice spacing $a$. The lowest band is filled by electrons of density $n$. An electric field $E$ applied along the $x$ direction. The electron scattering is negligible, the temperature is low, and the quasiclassical approximation is valid. An oscillatory electron response, called *Bloch oscillations*, is expected.

(*a*) Calculate the fundamental frequency of the oscillation for $a = 50$ Å and $E = 5 \times 10^4$ V/cm. Compare the result to the typical relaxation time of the electrons.

(*b*) Explain, in a few words, how does the oscillation in the electric polarization develop. Sketch the electric polarization vs. time for various band fillings.

# 9 Interactions and Phase Transitions

By definition, solids are made of strongly interacting atoms. One of the great successes of solid state theory is the understanding of many of the properties of solids in terms of a bunch of noninteracting quasiparticles – phonons, magnons, or electrons with no Coulomb repulsion. In the end, however, one has to face reality: Quasiparticles do interact with each other, even if the interaction may be weak. In this chapter we will discuss some of the consequences of these interactions.

If one looks at the ground state of a system, the effects of the interactions fall into two, broadly defined categories. As the strength of the interaction is increased from zero, the basic symmetries of the system may remain unchanged, but the numerical values of some parameters are modified (they get *renormalized*). The quasiparticles retain their basic character, but they acquire a finite lifetime. On the other hand, at sufficiently strong interaction, the ground state may suffer a symmetry change. Thinking in terms of the old quasiparticles become useless, and new ones are called for.

For a fixed strength of the interaction, the temperature dependencies of the properties of the material usually follow a similar pattern. The change of symmetry is called a *phase transition*.

Let us start with a discussion of the response of noninteracting conduction electrons to a time-dependent potential $\phi(r, t)$. Assuming that this perturbation is weak, the rearrangement of the charges, characterized by the induced charge density $\rho^{ind}$, is proportional to the potential. In terms of Fourier components,

$$\rho^{ind}(q, \omega) = \chi(q, \omega)\phi(q, \omega) , \tag{I.9.1}$$

where $\chi$, the susceptibility, remains to be determined. The simplest nontrivial result is based on second-order perturbation theory:

$$\chi = \frac{1}{V} \sum_k \frac{|\langle k + q | e^{-iqr} | k \rangle|^2 f(E(k + q)) - f(E(k))}{E(k + q) - E(k) - \hbar\omega + i\hbar\alpha} , \tag{I.9.2}$$

where $V$ is the volume of the system, $\langle k + q | e^{-iqr} | k \rangle$ is the matrix element of the electronic wavefunctions with a plane wave, $f(k) \equiv f(E(k))$ is the Fermi function, and the result should be evaluated in the $\alpha \to 0$ limit.

Due to the energy denominator in the sum, the largest contribution to occurs for $q$ vectors connecting nearly equal energy points on the energy surface. At the same time, the Fermi functions ensure that $q$ must connect empty states to filled states. Thus $\chi(q)$ will have sharp peaks, and the electronic response will be particularly strong, if there is a good match between the original Fermi surface and another one shifted by $q$. This match is called "Fermi surface nesting." The strong electronic response leads to *Kohn anomalies* in the phonon spectrum.[1]

What happens now if the electrons interact? In the spirit of the *mean field approximation* or *random phase approximation* (RPA) we assume that the $\phi(r)$ potential felt by any given electron is in fact composed of the externally applied potential ($\phi^{ext}$) and the potential created by the induced charge density.[2] Using the first Maxwell equation, $\nabla E = 4\pi\rho^{ind}$, the potential is expressed as $\phi = \phi^{ext} + U\rho^{ind}$, where $U = 4\pi e/q^2$ is the Fourier transform of the Coulomb interaction potential between the electrons; in a more general framework, we can take an arbitrary potential $U(q)$. The dielectric function $\epsilon(q,\omega)$ is

$$\epsilon(q,\omega) = \frac{\phi^{ext}(q,\omega)}{\phi(q,\omega)} . \tag{I.9.3}$$

The *renormalized susceptibility* is defined, similar to Eq. I.9.1, as

$$\rho^{ind}(q,\omega) = \chi^*(q,\omega)\phi^{ext}(q,\omega) . \tag{I.9.4}$$

In our approximation the dielectric function can be expressed as

$$\epsilon(q,\omega) = 1 - U(q)\chi(q,\omega) . \tag{I.9.5}$$

The renormalized susceptibility incorporates the effects of the interactions:

$$\chi^*(q,\omega) = \frac{\chi(q,\omega)}{1 - U(q)\chi(q,\omega)} . \tag{I.9.6}$$

To illustrate these concepts, let us now consider the nearly free-electrons, interacting with the Coulomb potential. The matrix element in Eq. I.9.2 is easily calculated, and the *Linhard formula*[3] is obtained:

$$\epsilon(q,\omega) = 1 - \frac{4\pi e^2}{q^2}\frac{1}{V}\sum_{k,\sigma}\frac{f(E(k+q)) - f(E(k))}{E(k+q) - E(k) - \hbar\omega + i\hbar\alpha} , \tag{I.9.7}$$

where $E(k) = \hbar^2 k^2/2m$ .

---

[1] See Ziman [3] p. 155, or Ashcroft and Mermin [1] p. 515.
[2] See, for example, Ziman [3] pp. 146–170, Ashcroft and Mermin [1] pp. 330–352, Kittel [2] pp. 296–317, Harrison [5] pp. 280–314, or Ibach and Lüth [4] pp. 97–99.
[3] See Eq. 5.16 of Ziman [3], Eq. 17.56 of Ashcroft and Mermin [1], Eq. 3.52 of Harrison [5].

For a static potential, assuming that the energy variation within a primitive unit cell of the crystal is small compared to the bandwidth, Eq. I.9.7 reduces to the *Thomas–Fermi formula*:

$$\epsilon(\boldsymbol{q}, \omega = 0) = 1 + \frac{k_0^2}{q^2} , \qquad (\text{I.9.8})$$

where $k_0 = \sqrt{4\pi e^2 g(E_F)}$. Consequently, the Coulomb potential of the probe charge $Q$, $\phi(x) = Q/r$, turns into a screened Yukawa potential $\phi(x) = (Q/r)e^{-k_0 r}$. The quantity $1/k_0$ is called the Thomas–Fermi screening length and its value is of the order of the lattice spacing. For lightly doped semiconductors with a very dilute electron gas (when classical statistics apply) a similar process is called the *Debye–Hückel* screening. The characteristic screening parameter is

$$k_0 = \sqrt{\frac{4\pi e^2 n}{k_B T}} , \qquad (\text{I.9.9})$$

where $n$ is the electron density. Here the screening length is much longer than the lattice spacing.

Eq. I.9.8 is attractive for its simplicity, but one has to keep in mind that it is an approximate formula only, not valid for large $q$. For example, an exact integration of Eq. I.9.7 for a three-dimensional free-electron band, $E(\boldsymbol{k}) = \hbar^2 k^2 / 2m$, leads to

$$
\begin{aligned}
\epsilon(\boldsymbol{q}, \omega = 0) &= 1 + \frac{4\pi m e^2 k_F}{\hbar^2 \pi^2 q^2} \left\{ \frac{1}{2} + \frac{1 - x^2}{4x} \ln \left| \frac{1 + x}{1 - x} \right| \right\} \\
&\equiv 1 + \frac{k_0^2}{q^2} L(x) , \qquad (\text{I.9.10})
\end{aligned}
$$

where $x = \frac{q}{2k_F}$. The function $L(x)$ is plotted in Figure I.9.1. The drop of $L(q/2k_F)$ for wavenumbers above $q = 2k_F$ indicates that the screening becomes less effective for rapid variations of the potential.

Another limitation of our result follows from the application of perturbation theory in calculating the response of the electrons. At small $q$ the energy denominator, $E(\boldsymbol{k} + \boldsymbol{q}) - E(\boldsymbol{k})$, becomes small, whereas the matrix element of the perturbation potential, $U(\boldsymbol{q}) \propto 1/q^2$, diverges. Therefore there is not much reason to believe that Eq. I.9.7 works for small $q$. In reality, the shielded electron–electron interaction is not a simple Yukawa potential. However, as $q \to 0$ the dielectric function remains divergent, indicating that the conduction electrons totally shield the external potential for large distances.

At the poles of the function $\epsilon^{-1}(\boldsymbol{q}, \omega)$, the internal potential may be nonzero even if there is no external potential. This follows from Eq. I.9.3: $\phi = \epsilon^{-1} \phi^{ext}$. In other words, when $\epsilon(\boldsymbol{q}, \omega)$ is zero, spontaneous oscillations develop in the system with wavenumber $\boldsymbol{q}$ at the frequency $\omega$.[4] Assuming

---

[4] This is similar to the result discussed in the introduction to chapter 8, Eq. I.8.10; for $\epsilon = 0$, propagating wave solutions of the Maxwell equation are possible.

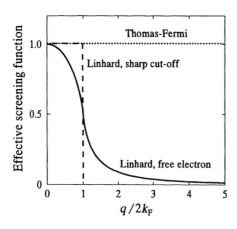

**Fig. I.9.1.** The wavenumber-dependent coefficient in the Lindhard function (solid line). The dotted line corresponds to the Thomas–Fermi approximation. The dashed line is another approximation, where the Linhard function is replaced by a function with a sharp cut-off at $2k_F$.

$\hbar\omega \gg |E(\boldsymbol{k} = \boldsymbol{q}) - E(\boldsymbol{k})|$, and expanding the terms in Eq. I.9.7 we obtain

$$\epsilon(\boldsymbol{q}, \omega) = 1 - \frac{1}{\omega^2}\omega_p^2 \left(1 + \frac{3}{5}\frac{q^2 v_F^2}{\omega_p^2} + ...\right), \tag{I.9.11}$$

where $v_F$ is the Fermi velocity, and $\omega_p^2 = 4\pi n e^2/m$. For $q = 0$ this result is identical to the one obtained in a much simpler classical model, Eq. I.8.24. The $\epsilon = 0$ condition, combined with Eq. I.9.11 provide us with the first term in the (weak) dispersion of the so called *plasma oscillations* or *plasmons*:

$$\omega = \omega_p \left(1 + \frac{3}{10}\frac{q^2 v_F^2}{\omega_p^2} + ...\right). \tag{I.9.12}$$

If $\epsilon(\boldsymbol{q}, \omega) = 0$ occurs at $\boldsymbol{q} = \boldsymbol{q}_0$ and at zero frequency, then a static $\phi(\boldsymbol{q}_0)$ potential develops. The renormalized susceptibility diverges and a density wave appears, even if no external potential is applied. A similar instability in the magnetic susceptibility indicates the presence of a Spin Density Wave (SDW). The concept of susceptibility can be further generalized, and superconductivity and ferromagnetism can also be discussed in the same framework. In all cases the divergence of the appropriate susceptibility indicates that a symmetry of the original crystal is spontaneously broken at low temperatures, a new phase develops, and the system is no longer metallic.

In the approximations we used (see Eq. I.9.6), $\epsilon(\boldsymbol{q}) = 0$ or $\chi^* \to \infty$ if $U\chi = 1$. For attractive interactions $\chi$ should be negative; for repulsive interactions it should be positive. (For nearly free-electrons the susceptibility is always negative; see Eq. I.9.10.) Even an arbitrarily weak interaction leads to a phase transition if the "bare" $\chi(\boldsymbol{q})$ diverges. At finite temperatures the Fermi functions in Eq. I.9.2 smooth out the divergence in $\chi$. In general, as the temperature is raised a second-order phase transition occurs and the metallic state is restored.

The tendency towards instability is over-emphasized by the mean field approximation. For example, as we will see in Problems 9.6 and 9.9, at temperatures below a critical temperature $T_c$, the one-dimensional metallic electron system is unstable: For an arbitrarily small attractive interaction, $U(q = 2k_F) < 0$, a Charge Density Wave (CDW) develops at wavenumbers $q = 2k_F$. However, in these calculations the fluctuations within the electronic system were totally neglected. The exact results for a one-dimensional electron system show that, strictly speaking, the static CDW only exists at $T = 0$, although long range, fluctuating CDW segments develop at $T < T_c$. In real materials a weak coupling between the one-dimensional chains results in a finite temperature phase transition.[5]

The *Mott transition* and *Wigner crystallization* are two more ways of looking at metal–insulator transitions due to the electron–electron interaction. Mott's argument starts with a material where the electronic band is partially filled, but the density of the atoms is low. Let us take, for example, one valence electron per atom. With no interaction between the electrons the material should be a metal (see Problem 3.12), even if the atoms are very far apart. However, common sense suggests that for a large enough separation between the atomic cores the valence electrons will not make a conduction band, but instead they will be bound to the atoms. According to Mott, a sharp transition from the metallic to the insulating state happens as the distance between the atoms exceeds $a_{crit} = \alpha \hbar^2/me^2$, where $\alpha$ is a number around 2 to 5. Wigner investigated the stability of interacting electrons assuming plane wave wavefunctions. He concluded that when bandwidth ($\approx \hbar^2 \pi^2/2ma^2$) is comparable to the strength of the interaction ($\approx e^2/a$), the "metallic" solutions with with uniform electron density become unstable (here the density of the electrons is $1/a^3$). In spite of the extreme opposite starting points (the tight-binding model for Mott and nearly free-electrons for Wigner), the conclusion is surprisingly similar: The metallic state does not survive at low conduction electron densities.[6]

## 9.1 Problem: Spontaneous Polarization

The polarization of a molecule is described by the anharmonic equation of motion,

$$\frac{\partial^2 p}{\partial t^2} = -\omega_0^2 p - \lambda p^3 + E\alpha\omega_0^2 . \tag{I.9.13}$$

---

[5] Charge density waves have been observed in quasi-one-dimensional materials like $TaS_3$, $NbSe_3$ or organic conductors like TTF-TCNQ (tetramethyl tetrathio fulvalene - tetracyano quinodimethane). Spin density waves occur in (TMTSF) $PF_6$. For a detailed discussion see Grüner [16].

[6] *Anderson localization* is another way of destroying the metallic ground state. It happens in disordered metals if the bandwidth is comparable to the random potential due to impurities. The Mott and Anderson processes have different physical origins, but they both lead to the same result: The elimination of the metallic state of partially filled narrow bands.

A linear chain of lattice spacing $a$ is made of these molecules. The polarization is parallel to the chain, and each molecule is exposed to the electric field of the others. The temperature is zero, and quantum effects are negligible.

(a) For fixed values of the other parameters and $\alpha > \alpha_{crit}$ the system develops a spontaneous polarization. What is the critical value of $\alpha$?

(b) Calculate the magnitude of the static polarization $p_0$ as a function of $\alpha$. Show that, close to $\alpha_{crit}$, $p_0 \sim (\alpha - \alpha_{crit})^\beta$, and calculate the exponent $\beta$.

## 9.2 Problem: Divergent Susceptibility

Consider a linear chain of molecules. The polarization $p$ of each molecule is described by the classical equation of motion,

$$\frac{\partial^2 p}{\partial t^2} = -\omega_0^2 p + E\alpha\omega_0^2 , \tag{I.9.14}$$

where $E$ is the local electric field. The polarization is aligned along the chain, and each molecule is exposed to the electric field generated by the others. The $p = 0$ ground state of this system becomes unstable if the polarizability exceeds[7] $\alpha_{crit} = a^3/4.808\omega_0^2$.

The susceptibility characterizes the response to an external perturbation. The relevant perturbation for this system is an external electric field wave, $E = E^* e^{ikx-\omega t}$. The response is a polarization wave, $p = p^* e^{ikx-\omega t}$. The susceptibility is defined as $p^* = \chi(k,\omega)E^*$.

(a) Calculate the susceptibility as a function of the wavenumber $k$ at $\omega = 0$ for $\alpha < \alpha_{crit}$. Show that $\chi(k = 0, \omega = 0)$ diverges at $\alpha \leq \alpha_{crit}$ as $\chi \sim (\alpha_{crit} - \alpha)^\gamma$, and calculate the exponent of the divergence, $\gamma$.

(b) Calculate $\chi(k)$ at $\alpha = \alpha_{crit}$,

## 9.3 Problem: Large-U Hubbard Model

The Hubbard model Hamiltonian,

$$H = \sum_{i \neq j, \sigma}^{N} t_{ij} a_{i,\sigma}^+ a_{j,\sigma} + \sum_i^{N} U\, n_{i\uparrow} n_{i\downarrow}, \tag{I.9.15}$$

is used for delocalized electrons interacting with an on-site interaction $U$ ($n_i = a_i^+ a_i$ is the number of electrons at site $i$; the spin $\sigma$ is represented by the up and down arrows, the number of electrons is $N$.) Typically, $t_{ij} = t$ for

---

[7] See Problems 2.16 and 9.1.

first neighbors and $t_{ij} = 0$ otherwise (see Problem 5.1). Electrons with the same spin cannot be on the same site.

The spin Hamiltonian $H = \frac{1}{2}\sum_{i \neq j} J_{ij}S_iS_j$ describes a system of localized electrons (Heisenberg model). We will take $J_{ij} = J$ for first neighbors and $J_{ij} = 0$ otherwise. For a half-filled electronic band and strong repulsive interaction ($U > 0$, and $U \gg t$), the low-energy excitations of the electronic system match closely the excitations of the magnetic system.

The proof of this theorem is rather complicated for large $N$. Show that the equivalence is true for $N = 2$, that is for systems consisting of two electrons (two spins). Calculate $J$ for given $U$ and $t$.

## 9.4 Problem: Infinite Range Hubbard Model

Interacting electrons are described by a Hamiltonian

$$H = \sum_{k,\sigma} E(k)a^+_{k,\sigma}a_{k,\sigma} + \frac{U}{V}\sum_{k,k'} a^+_{k,\uparrow}a^+_{k',\downarrow}a_{k',\downarrow}a_{k,\uparrow} , \qquad (I.9.16)$$

where $\sigma = \uparrow$ or $\downarrow$ represents the electron spin, $k$ and $k'$ are electron wavenumbers, and $V$ is the volume of the system.[8] For $U = 0$ the elementary excitations are created by the operator $a^+_{k,\sigma}$. The excitation spectrum is $E(k)$. The density of particles in spin state $\sigma$, $n_\sigma = (1/V)\sum_k a^+_{k,\sigma}a_{k,\sigma}$, commutes with $H$, therefore it is a good quantum number of the system.

Show that, for arbitrary values of $U$, the elementary excitations are still created by $a^+_{k,\sigma}$, and $n_\sigma$ is still a good quantum number. Determine the excitation spectrum $E(k)$, in terms of $U$, $n_\uparrow$, and $n_\downarrow$.

## 9.5 Problem: Stoner Model

Calculate the low temperature magnetic susceptibility, $\chi$, of the system described by the Hamiltonian

$$H = \sum_{k,\sigma} E(k)n_{k,\sigma} + \frac{U}{V}\sum_{k,k'} n_{k,\uparrow}n_{k',\downarrow} , \qquad (I.9.17)$$

where $\sigma = \uparrow$ or $\downarrow$ represents the electron spin, $k$ and $k'$ are electron wavenumbers, $n_{k,\sigma}$ is the electron number operator, and $V$ is the volume of the system.[9] Assume that the band structure $E(k)$, and the corresponding density of states $g(E)$, are known. Discuss $\chi$ as a function of the interaction parameter $U$.

---

[8] This Hamiltonian is used in the *Stoner model* of conduction electron ferromagnetism. (See Ziman [3] p. 339). In terms of the more general, wavenumber dependent potential $U(q)$, the second term in the Hamiltonian corresponds to $U(q) = U\delta(q)$. A Fourier transformation reveals that this represents an "infinite range" interaction between the electrons.

[9] This Hamiltonian is equivalent to the one in Problem 9.4.

## 9.6 Problem: One-Dimensional Electron System

The static dielectric function for an interacting electron system is

$$\epsilon(q) = 1 - U(q)\frac{1}{V}\sum_{k,\sigma}\frac{f(k+q) - f(k)}{E(k+q) - E(k)}, \qquad (I.9.18)$$

where $U(q)$ is the interaction potential, $V$ is the volume and $f(k) = f(E(k))$ is the Fermi function. Consider a one-dimensional system, where the energy depends on $k_x$, but is independent of $k_y$ and $k_z$. The energy is given by $E = \hbar^2 k_x/2m$, and the Fermi surface is at $k_x = \pm k_F$. Investigate the properties of the $\epsilon(q = (q_x, 0, 0))$ function at low temperature. Show that $\epsilon$ becomes zero at $q_x = 2k_F$ for an arbitrarily small repulsive interaction, $U(q = 2k_F) > 0$.

## 9.7 Problem: Peierls Distortion

In a simple model of the coupled electron–phonon system the lattice energy is assumed to be $E_{ph} = \frac{1}{2}\kappa s^2$, where $s$ is the amplitude of a static lattice distortion of wavevector $q$, and $\kappa$ is an elastic constant of the crystal. The static distortion of the lattice acts on the electrons as a perturbation, which is described by a weak periodic potential, $U = U_0 \cos qx$, where $U_0^2 = \alpha s^2$ and $\alpha$ is a coupling constant describing how sensitive the electrons are to a lattice distortion.

The periodic potential will induce a gap in the electronic spectrum. If the wavenumber of the modulation is $q = 2k_F$, then the energy gap will open at $\mathcal{E}_F$, and the Fermi energy will be in the middle of the band gap. Assuming that the perturbation is much less than the bandwidth, $U_0 \ll W$, the density of states around $\mathcal{E} = \mathcal{E}_F$ can be approximated as

$$g(\mathcal{E}) = g_0 \frac{\mathcal{E} - \mathcal{E}_F}{\sqrt{(\mathcal{E} - \mathcal{E}_F)^2 - \Delta^2}}, \qquad (I.9.19)$$

where $g_0$ is the density of states for the nonperturbed system and $\Delta = U_0/2$.[10]

(a) Calculate the electronic contribution to the free energy $k_B T \ll \Delta$ and for $k_B T \gg \Delta$, and discuss if the free energy increases or decreases when the gap opens.

(b) Investigate the stability of the metallic state ($\Delta = 0$) of the coupled electron–phonon system at low temperatures. Determine the magnitude of the energy gap.

(c) Calculate the temperature of the *Peierls transition*, when a spontaneous lattice distortion develops in the system.

---

[10] See Solution 4.9.

## 9.8 Problem: Singularity at $2k_F$

Calculate the Linhard function,

$$L(q) = \frac{\chi}{g(E_F)} = \frac{1}{g(E_F)} \int d^d k \frac{f(k+q) - f(k)}{E(k+q) - E(k)} , \qquad (I.9.20)$$

for a $d = 1$-, 2-, and 3-dimensional electron gas in the low temperature limit. (Here $f$ is the Fermi function, $E(k) = \hbar^2 k^2/2m$, and $g(E_F)$ is the density of states at the Fermi level.)

## 9.9 Problem: Susceptibility of a One-Dimensional Electron Gas

At zero temperature the susceptibility $\chi(q)$ of a one-dimensional electron gas diverges at $q = 2k_F$. Determine the temperature dependence of $\chi(q = 2k_F)$ in the $T \to 0$ limit.

## 9.10 Problem: Critical Temperature in Mean Field Approximation

In the mean field approximation an electron system with an attractive interaction $(U < 0)$ becomes unstable at low temperatures. The critical temperature $T_c$ is defined by the condition $\epsilon \equiv 1 - U\chi = 0$, where $\chi$ is the susceptibility. Calculate $T_c$ in the weak coupling limit, $U \ll E_F$. Show that $k_B T_c = \beta E_F e^{-1/Ug}$, where $g$ is the density of states at the Fermi level and $\beta$ is a constant of the order of 1.

## 9.11 Problem: Instability of Half-Filled Band

For electrons interacting by an interaction potential $U(q)$, the static dielectric function can be expressed as $\epsilon(q) = 1 - U(q)\chi(q)$. Let us approximate the susceptibility by

$$\chi(q) \approx \frac{1}{V} \sum_{k,\sigma} \frac{f(k+q) - f(k)}{E(k+q) - E(k)} \equiv -G(q) , \qquad (I.9.21)$$

where $V$ is the volume and $f(k) = f(E(k))$ is the Fermi function.[11]

---

[11] The more accurate expression for the susceptibility of the noninteracting electron gas (see Eq. I.9.2) involves the $\langle k + q | e^{-iqr} | k \rangle$ matrix element, which is exactly 1 for free electrons, but it has a momentum dependence for other electron wavefunctions. However, the momentum dependence is smooth, and the main features of the result do not change if we replace the matrix element by a constant of the order of one.

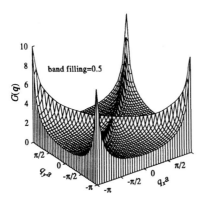

**Fig. I.9.2.** The Fermi surface integral $G(q)$, calculated for a half-filled band. The temperature was set to a small but nonzero value to eliminate the divergences.

In Figure I.9.2 the $G(q)$ function is plotted for a half-filled tight-binding electron system on a square lattice. The energy band is described by $E = -E_0(\cos k_x a + \cos k_y a)/4$. Show that at low temperatures the $G(q)$ function diverges at the $q = (\pm\pi/a, \pm\pi/a)$ points. (If $U < 0$, the susceptibility is divergent, and the dielectric function becomes zero for an arbitrarily small $U$. This result indicates that an attractive interaction can produce charge density waves at low temperatures.)

## 9.12 Problem: Screening of an Impurity Charge

In a three-dimensional free-electron system the dielectric function is described by the Linhard function, Eq. I.9.7. Take a simple approximation for the effective screening function $L(q/2k_F)$: $L = 1$ for $q < 2k_F$, and $L = 0$ otherwise (Figure I.9.1). Use this function to investigate the asymptotic behavior of the electric potential $\phi(r)$ around a point charge impurity.[12]

## 9.13 Problem: Fermi Surface Nesting in Two Dimensions

In Figure I.9.3, the

$$G(q) = -\frac{1}{V}\sum_{k,\sigma}\frac{f(k+q) - f(k)}{E(k+q) - E(k)} , \qquad (I.9.22)$$

function is plotted for a two-dimensional electron system on a square lattice, characterized by the energy dispersion $E = E_0(-\cos k_x a - \cos k_y a + 2)/4$ . The band is nearly half-filled, $E_F = 0.4E_0$, corresponding to band filling of 0.42. The temperature is much less than the Fermi temperature. Determine

---

[12] This is a *Friedel oscillation*, as discussed by Ziman [3] p. 159, and others.

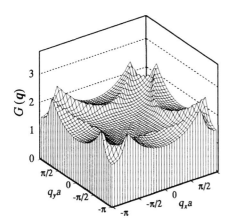

**Fig. I.9.3.** The Fermi surface integral $G(q)$ for a nearly half-filled tight-binding band.

the peak positions in $G(q)$ as a function of $\eta = 1/2 - E_F/E_0$ for small $\eta$.[13]

## 9.14 Problem: Fermi Surface Nesting in Quasi One Dimension

A quasi-one-dimensional electron system is described by

$$E = -E_0(\cos k_x a + \alpha \cos k_y b) , \qquad (\text{I}.9.23)$$

where $\alpha \ll 1$. The band is partially filled so that the Fermi lines are open, as indicated in Figure I.9.4. Due to "Fermi surface nesting", at low temperatures the susceptibility $\chi(q)$ develops peaks at certain values of $q = q_0$. Determine the peak position $q_0$ as a function of $\alpha$ for $\alpha \ll 1$.

## 9.15 Problem: Anderson Model

The Anderson Hamiltonian[14] describes electrons in a transition metal, including the interactions between the electrons:

$$
\begin{aligned}
H \; = \; & \sum_k E_c(k)(n_{k,\uparrow} + n_{k,\downarrow}) + \sum_d E_d(n_{d,\uparrow} + n_{d,\downarrow}) \\
& - \Delta \sum_{k,d}(c^+_{k,\uparrow}d_\uparrow + d^+_\uparrow c_{k,\uparrow} + c^+_{k,\downarrow}d_\downarrow + d^+_\downarrow c_{k,\downarrow}) + U n_{d,\uparrow}n_{d,\downarrow} \quad (\text{I}.9.24)
\end{aligned}
$$

---

[13] As discussed in problem 9.11, the divergence of $G(q)$ indicates divergent susceptibility. At these wavenumbers Kohn anomalies occur in the phonon spectrum, or a weak electron–electron interaction will cause an instability of the metallic state.

[14] P.W. Anderson, *Phys. Rev.*, **124**, 41 (1961).

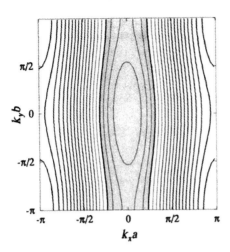

**Fig. I.9.4.** Contour plot of the energy surface for a quasi-one-dimensional electron system described by Eq. I.9.23, for $\alpha = 0.1$. The occupied electronic states are shaded, and the thick solid lines represent the Fermi lines.

where $\uparrow$ and $\downarrow$ indicate the spin state, $n_{k,\uparrow} = c_{k,\uparrow}^+ c_{k,\uparrow}$, and $n_{d,\uparrow} = d_\uparrow^+ d_\uparrow$; this is similar for "down" spins. The first term in the Hamiltonian corresponds to electrons in a broad conduction band with energy $E_c(k)$, where $k$ is the wavenumber. Assume that $E_c(k)$ is known and that the corresponding density of states, $g_c(k) = g_0$, is constant. The second and third terms describe electrons in the narrow $d$ band and the transfer between the two bands, respectively. There are $N_d$ states in the narrow band for each spin. The last term represents the interaction between the electrons on the $d$ levels; only electrons with opposite spins are allowed to occupy these levels.

The bandwidth corresponding to $E_c(k)$ is much larger than the transfer matrix element $|\Delta|$. The Fermi energy $E_F$, $E_d$, and $E_d + U$ are all well within the conduction band.[15]

(a) What kind of magnetic behavior is expected for $\Delta = 0$, if the number of $d$ electrons is $N_d^e = N_d$? Discuss what happens to the magnetic properties if $E_d$ is less than or greater than $E_F$, and $E_d + U$ is less or greater than $E_F$.

(b) For $\Delta > 0$, the exact and full solution to the problem is not known. However, within the framework of the *mean field* approximation, we can have some ideas about the behavior of the system. Replace the Anderson Hamiltonian with

$$
\begin{aligned}
H_\uparrow = {} & \sum_k E_c(k) n_{k,\uparrow} + \sum_d E_d n_{d,\uparrow} \\
& - \Delta \sum_{k,d} (c_{k,\uparrow}^+ d_\uparrow + d_\uparrow^+ c_{k,\uparrow}) + U \langle n_{d,\downarrow} \rangle n_{d,\uparrow} ,
\end{aligned} \tag{I.9.25}
$$

---

[15] The first three terms of Eq. I.9.24 describe a system similar to that investigated in Problem 5.2.

and a similar one for the ↓ spins. Here $\langle n_{d,\downarrow} \rangle = \frac{1}{N_d} \sum_{k,d} \langle \psi_{k,d} | d_\uparrow^\dagger d_\downarrow | \psi_{k,d} \rangle$ is the expectation value of the ↓ spin occupation of the $d$ state (the sum is over the occupied states).

Use the self-consistency condition to calculate the magnitude of the localized moment, $\mu = \mu_B (\langle n_{d,\uparrow} \rangle - \langle n_{d,\downarrow} \rangle)$ as a function of $(E_F - E_d)/\delta$ and $U/\delta$, where $\delta = \Delta^2 g_0 / N_d$.

*Hint*: Search for the solution in the form of $|\psi_k\rangle = (\alpha_k c_k^+ + \beta_k d^+)$ with the normalization condition $|\alpha_k|^2 + |\beta_k|^2 = 1$. The expectation value $\langle n_{d,\uparrow} \rangle = \sum_{k,d} |\beta|^2$ can be converted to an energy integral: $\langle n_{d,\uparrow} \rangle = \int_{-\infty}^{E_F} g(E) \beta(E)^2 \, dE$, where $g(E)$ is the density of states, determined by solving Eq. I.9.25.

# PART II
## Solutions to Problems

# 1 Crystal Structures

## 1.1 Solution: Symmetries

Let us first look at the crystal in Figure I.1.5b. There are eight elements in the symmetry group: The identity operation (when no points move), 90°, 180°, and 270° rotations, and four mirror lines, as indicated in Figure II.1.1. In two dimensions the inversion operation is not a separate symmetry, since it is equivalent to the 180° rotation.

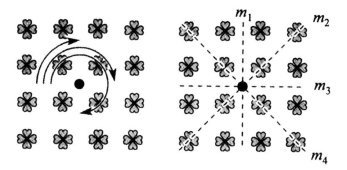

**Fig. II.1.1.** Symmetry operations of the crystal shown in Figure I.1.5b. The lines $m_1$, $m_2$, $m_3$, and $m_4$ are mirror lines.

The polygon has the same symmetries. The simplest object with this symmetry group is a square; the point group is denoted by $4mm$.

The crystal in Figure I.1.5a does not have mirror lines, and therefore the number of elements in the symmetry group is four. If the two groups have different numbers of elements, they must be different symmetry groups.

## 1.2 Solution: Rotations

Applying the usual rules of matrix multiplication one obtains: $AA = B$, $AB = E$, and so on. The multiplication table is then

|   | E | A | B |
|---|---|---|---|
| E | E | A | B |
| A | A | B | E |
| B | B | E | A |

$$(II.1.1)$$

A point group of three elements suggests that the crystal may have a three-fold rotational symmetry and nothing else. A quick check of the symmetry operations allows us to conclude that the point group of the crystal in Figure I.1.5a has a multiplication table matching the one shown above.

## 1.3 Solution: Copper Oxide Layers

(a) One of many choices for the Bravais lattice is shown in Figure II.1.2 along with its unit cell, basis, and primitive vectors.

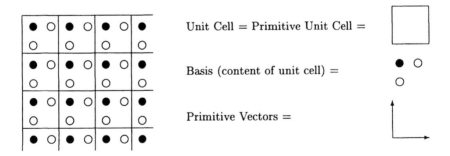

Unit Cell = Primitive Unit Cell =

Basis (content of unit cell) =

Primitive Vectors =

**Fig. II.1.2** CuO$_2$ Bravais lattice.

(b) A choice for the Bravais lattice in this case, as well as for the new primitive unit cell, is shown in Figure II.1.3. Note the new lattice spacing is $\sqrt{2}a$.

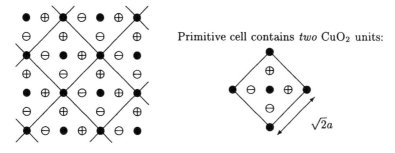

**Fig. II.1.3** Distorted $CuO_2$ Bravais lattice.

The reciprocal lattices for both Bravais lattices are shown in Figure II.1.4, the open circles correspond to weak X-ray diffraction peaks, and the filled circles are strong peaks. As the distortion decreases, the structure factor of the weaker Bragg peaks of the distorted crystal will go to zero. Hence the X-ray pattern will return to the original reciprocal lattice pattern.

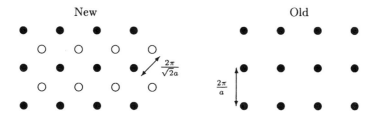

**Fig. II.1.4.** Reciprocal lattices for distorted (new) and pristine (old) $CuO_2$ lattices.

## 1.4 Solution: Graphite

Graphite has four atoms in a primitive unit cell. Each honeycomb layer of carbon atoms needs two atoms per cell. The subsequent layers are shifted by a half-lattice spacing, leading to another factor of two.[1]

## 1.5 Solution: Structure of $A_x C_{60}$

(a) Figure 2a of Fleming *et al.* shows an unconventional unit cell for the $fcc$ structured $A_3 C_{60}$ solid. Transforming into a conventional $fcc$ unit cell can be accomplished by taking an axis along the smaller face diagonal of Fleming's

---

[1] See Ashcroft and Mermin [1] p. 304, or Ziman [14] p. 120.

cell. This is sketched in Figure II.1.5. We see that the $C_{60}$ molecule in the center of Fleming's cell becomes a face center in the conventional unit cell; using the axes defined in Figure II.1.5, it's coordinates would be $(\frac{1}{2}, 0, \frac{1}{2})$.

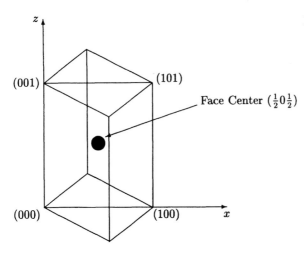

**Fig. II.1.5.** Conventional *fcc* unit cell face drawn on Fleming's $A_3C_{60}$ unit cell.

(b) $Rb_x C_{60}$ has an *fcc* structure for $x = 3$, *bct* for $x = 4$, and *bcc* for $x = 6$. To calculate the X-ray peak positions, $2\theta$, we must use

$$G = \sqrt{(h \cdot b_1)^2 + (k \cdot b_2)^2 + (l \cdot b_3)^2} \,, \tag{II.1.2}$$

where $b_1 = b_2 = \frac{2\pi}{a}$, and $b_3 = \frac{2\pi}{a}$ for *fcc* and *bcc* or $b_3 = \frac{2\pi}{c}$ for *bct*. We can then use the condition $2\frac{2\pi}{\lambda} \sin\theta = G$ to calculate values of $2\theta$.

Since we are using the conventional cubic unit cells, we must calculate the structure factor, Eq. I.1.2, to determine which $(h, k, l)$ values are allowed.

$$S_K = \sum_{j=1}^{n} e^{i K \cdot d_j} \tag{II.1.3}$$

for $n$ scatterers at positions $d_j$ in a unit cell.

For the *fcc* unit cell, there is a $C_{60}$ at position $d_1 = 0$, and on the three faces $d_2 = \frac{a}{2}(\hat{x} + \hat{y})$, $d_3 = \frac{a}{2}(\hat{x} + \hat{z})$, $d_4 = \frac{a}{2}(\hat{y} + \hat{z})$. The general vector is $K = \frac{2\pi}{a}(h\hat{x} + k\hat{y} + l\hat{z})$. Therefore

$$
\begin{aligned}
S_K &= 1 + e^{i\pi(h+k)} + e^{i\pi(h+l)} + e^{i\pi(k+l)} & \text{(II.1.4)}\\
&= 1 + (-1)^{h+k} + (-1)^{h+l} + (-1)^{k+l} \,. & \text{(II.1.5)}
\end{aligned}
$$

When $S_K \neq 0$, the Bragg peak associated with the reciprocal lattice vector $K$ at scattering angle $2\theta$ is allowed.

Similarly we can look at the two-molecule per unit cell *bct* and *bcc* structures to determine that their structure factors are given by

$$S_K = 1 + (-1)^{h+k+l}. \tag{II.1.6}$$

Tables II.1.1–II.1.3 show the resultant calculated values for the first five allowed diffraction peaks for the three structures.

**Table II.1.1.** First Five Diffraction Peaks for an *fcc* $Rb_3C_{60}$ Crystal, with $\lambda = 0.9$, $a = 14.436$ Å

| h | k | l | G | $2\theta$ |
|---|---|---|---|---|
| 1 | 1 | 1 | 0.7538651 | 6.189992 |
| 2 | 0 | 0 | 0.8704885 | 7.148748 |
| 2 | 2 | 0 | 1.231057 | 10.11644 |
| 3 | 1 | 1 | 1.443542 | 11.86838 |
| 2 | 2 | 2 | 1.50773 | 12.39814 |

**Table II.1.2.** First Five Diffraction Peaks for a *bct* $Rb_4C_{60}$ Crystal, with $\lambda = 0.9$, $a = 11.962$ Å, and $c = 10.022$ Å

| h | k | l | G | $2\theta$ |
|---|---|---|---|---|
| 1 | 1 | 0 | 0.7428328 | 6.09932 |
| 1 | 0 | 1 | 0.8178955 | 6.716326 |
| 2 | 0 | 0 | 1.050524 | 8.629824 |
| 0 | 0 | 2 | 1.253878 | 10.30448 |
| 2 | 1 | 1 | 1.331373 | 10.94322 |

**Table II.1.3.** First Five Diffraction Peaks for a *bcc* $Rb_6C_{60}$ Crystal, with $\lambda = 0.9$, $a = 11.548$ Å

| h | k | l | G | $2\theta$ |
|---|---|---|---|---|
| 1 | 1 | 0 | 0.7694637 | 6.318201 |
| 2 | 0 | 0 | 1.088186 | 8.939824 |
| 2 | 1 | 1 | 1.33275 | 10.95458 |
| 2 | 2 | 0 | 1.538927 | 12.65571 |
| 3 | 1 | 0 | 1.720573 | 14.15675 |

# 1.6 Solution: *hcp* and *fcc* Structures

The *fcc* structure is close packed, and so is the *hcp* structure with an ideal $c/a$ ratio. For $\alpha$-Co, $c/a = 1.62$, close to the ideal value of 1.63. In the *fcc* structure the nearest-neighbor distance is $3.55/\sqrt{2} = 2.51$ Å. The in-plane distance in the *hcp* structure is the same, $a = 2.51$ Å. If $c/a$ was 1.63, both structures would be closed packed, with the same nearest-neighbor distances and therefore same densities. With $c/a = 1.62$ the *hcp* structure is 0.6%

denser. This number is barely more than the accuracy of the numbers given in the problem.

## 1.7 Solution: *hcp* and *bcc* Structures

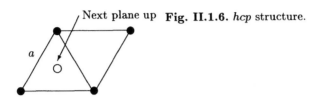

Next plane up  **Fig. II.1.6.** *hcp* structure.

The density of a *bcc* structure is 2 atoms per $(a')^3$; the density of an *hcp* structure is 2 atoms per the volume of the primitive unit cell, $V_0$. A sketch of the hexagonal closed packed structure is shown in Figure II.1.6. Three neighboring atoms in one plane, along with their neighbor in the next plane, form a regular tetrahedron with side $a$ and height $c/2$. The volume of the unit cell is $V_0 = \frac{\sqrt{3}}{2}ca^2$. If the furthest left atom in Figure II.1.6 is defined to be at $(0,0,0)$, then the out-of-plane atom is at coordinates $(\frac{a}{2}, \frac{a}{2\sqrt{3}}, \frac{c}{2})$. Since the distance between all atoms is $a$, we obtain the relation

$$ a = \sqrt{\frac{a^2}{4} + \frac{a^2}{12} + \frac{c^2}{4}} = \sqrt{\frac{a^2}{3} + \frac{c^2}{4}} \ . \tag{II.1.7} $$

Therefore the $c = \sqrt{8/3}a = 1.633a$ and $V_0 = \frac{\sqrt{8}}{2}a^3$. Solving for $a$ in terms of $a'$, we obtain $a^3 = (a')^3/\sqrt{2}$. With $a' = 4.23$ Å, the *hcp* lattice spacing is $a = 3.77$ Å.

## 1.8 Solution: Structure Factor of $A_x C_{60}$

The charge distribution on a buckyball is given by

$$ \rho(r) = \delta(r - R) \cdot A \ , \tag{II.1.8} $$

where $R = 3.5$ Å and $r = |\,r\,|$. Each carbon atom has 6 electrons. The total charge per $C_{60}$ is $6 \times 60 = 360$ electrons. Therefore

$$ 360e = \int \rho(r) \mathrm{d}^3 r = \int_0^\infty 4\pi r^2 \delta(r - R) \cdot A \ \mathrm{d}r \ , \tag{II.1.9} $$

and we solve for A resulting in $A = 360e/(4\pi R^2)$.

Using Eq. I.1.3, the form factor for this charge distribution is

$$
\begin{aligned}
f(\boldsymbol{K}) &= \frac{1}{e} \int \rho(\boldsymbol{r}) e^{i\boldsymbol{K}\cdot\boldsymbol{r}} d^3 r \\
&= \frac{1}{e} \int r^2 dr \delta(r-R) \frac{360e}{4\pi R^2} \int_0^{2\pi} d\phi \int_0^{\pi} e^{iKr\cos\theta} \sin\theta d\theta \\
&= \frac{360}{4\pi} 2\pi \int_1^{-1} (-1) d\cos\theta e^{iKR\cos\theta} \qquad\qquad (\text{II}.1.10) \\
&= 360 \frac{\sin KR}{KR} .
\end{aligned}
$$

Note that this is an oscillating function with its amplitude decaying for higher $K$ values.

For the (2 0 0) reflection the absolute value of the wavenumber is $K = \frac{2\pi}{a} \cdot 2 = 4\pi/14.11$ Å The form factor is

$$
f \sim \sin KR = \sin \frac{4\pi}{14.11 \text{ Å}} 3.5 \text{ Å} = 0.026 . \qquad (\text{II}.1.11)
$$

Due to this conspiracy of the numbers, the form factor for the (2 0 0) diffraction peak is very close to zero. The (1 1 1) reflection has $K = 2\sqrt{3}\pi/a$ and $\sin KR = 0.429$, resulting in a much larger form factor.[2]

## 1.9 Solution: Neutron Diffraction Device

If a beam of wavelength $\lambda$ is incident at an angle $\theta$ to a symmetry axis of a crystal, the Bragg scattering condition states

$$
n\lambda = 2b \sin \theta, \qquad\qquad (\text{II}.1.12)
$$

where $n$ is an integer and $b$ is the atomic spacing in the unit cell.

In a powder, the incident angle on each randomly oriented crystallite will be random between $0 < \theta < 2\pi$. Neutrons of wavelengths $\lambda \le 2b$ will be scattered at random angles out of the collimated beam while neutrons with longer wavelengths (lower energies) will pass through unaltered. This is an energy low-pass filter for neutron beams which can be used to restrict the energy of neutrons which are to be used in an experiment.

Neutron scatterers indeed use this effect with powders of beryllium. This material has an *hcp* structure, but the resultant low-pass filter has the same origins. In this case, neutrons with wavelengths of $\lambda > 3.96$ Å ($E < 5.2$ meV) will pass through. In practice the beryllium powder is also cooled to 77 K to decrease the scattering by thermal phonons. By eliminating higher-energy neutrons, the experimenter does not have to worry about "higher-order scattering" – where neutrons with double or triple the intended energy scatter unwanted into the detector, thereby artificially enhancing a signal.

---

[2] We are indebted to P.W. Stephens for many interesting discussions on X-ray diffraction.

## 1.10 Solution: Linear Array of Emitters, Finite Size Effects

In this problem, the nature of the radiation is not relevant (it could be electromagnetic waves, water waves, sound waves ...), but let us use the notations of electromagnetic radiation. The field at the detector will be the sum of the fields from each radiation source,

$$E = E_0 \sum_{n=1}^{N} e^{ikx_n} e^{i\omega t} , \qquad (\text{II.1.13})$$

where $N = L/a$ and $E_0$ is related to the amplitude of the sources and the distance between the detector and the sources. Without losing any important information we can assume that $E_0$ is real and independent of the distance to the sources.

The time-averaged intensity at the detector is

$$I = I_0 \left| \sum_{n=1}^{N} e^{i2\pi n\kappa} \right|^2 , \qquad (\text{II.1.14})$$

where $I_0 \sim E_0^2$ and we have defined $\kappa = \frac{k}{g}$, $g = \frac{2\pi}{a}$, and $N = \frac{L}{a}$, the total number of sources. Using the well-known formula for geometrical series, we obtain

$$\sum_{n=1}^{N} e^{i2\pi n\kappa} = \frac{e^{i2\pi N\kappa} - 1}{1 - e^{-i2\pi\kappa}} . \qquad (\text{II.1.15})$$

Note that for $\kappa = \nu$ (an integer), $k = \nu g$ and we get $e^{i2\pi n\kappa} = 1$ for each $n$. Therefore the sum in Eq. II.1.15 is equal to $N$ for $k = \nu g$.

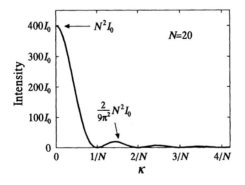

**Fig. II.1.7.** Intensity at the detector as a function of $\kappa$ for a linear array of emitters.

Inserting Eq. II.1.15 into Eq. II.1.14, we obtain

$$I = I_0 \left| \frac{e^{i2\pi N\kappa} - 1}{1 - e^{-i2\pi\kappa}} \right|^2 = I_0 \frac{1 - \cos(2\pi N\kappa)}{1 - \cos(2\pi\kappa)}. \qquad (\text{II.1.16})$$

This function is sketched in Figure II.1.7 (the same function as for light passing through a diffraction grating). The half-width of the peaks is $\kappa \sim \frac{1}{2N}$, or in terms of wavenumbers $k \sim \frac{g}{2N} = \frac{\pi}{L}$. The peak intensity is $N^2 I_0$, and the integrated intensity is approximately given by the half-width $\times$ height $\sim I_0 N$.

## 1.11 Solution: Linear Array of Emitters, Superlattice

Equation II.1.14 is modified to include the modulation,

$$I = I_0 \left| \sum_{n=1}^{\infty} e^{i2\pi n\kappa} \cdot e^{i2\pi\kappa u_n/a} \right|^2 \tag{II.1.17}$$

$$= I_0 \sum_{n,m} e^{i2\pi(n-m)\kappa} \cdot e^{i2\pi\kappa(u_n - u_m)/a} . \tag{II.1.18}$$

A series expansion of the second term yields

$$I = I_0 \sum_{n,m} e^{i2\pi(n-m)\kappa} \left[ 1 + i\frac{2\pi\kappa}{a}(u_n - u_m) \right.$$
$$\left. - \frac{1}{2} \left\{ \frac{2\pi\kappa}{a}(u_n - u_m) \right\}^2 + \cdots \right] . \tag{II.1.19}$$

Insert $u_n = u_0 \cos(nqa) = \frac{u_0}{2}(e^{inqa} + e^{-inqa})$ into Eq. II.1.19 and we obtain

$$I = I_0 \sum_{n,m} e^{i2\pi(n-m)\kappa} \left[ 1 + i\frac{\pi\kappa u_0}{a}(e^{iqna} + e^{-iqna} - e^{iqma} - e^{-iqma}) \right.$$
$$\left. - \frac{1}{2}\left(\frac{\pi\kappa u_0}{a}\right)^2 \{e^{iqna} + e^{-iqna} - e^{iqma} - e^{-iqma}\}^2 + \cdots \right], \tag{II.1.20}$$

where the term in curly brackets can be expanded as

$$\{\ \}^2 = 1 - e^{iq(n+m)a} - e^{iq(n-m)a} - e^{iq(m-n)a} - e^{iq(n+m)a}$$
$$+1 + e^{iq2na} + e^{iq2ma} + e^{-iq2na} + e^{-iq2ma} . \tag{II.1.21}$$

When the sum in Eq. II.1.20 is calculated, typical terms will look like

$$\sum_{n,m} e^{i2\pi\kappa'} \cdot e^{i2\pi\kappa''m} . \tag{II.1.22}$$

According to the geometrical series formula, the sum will be of order 1, except for integer values of $\kappa'$, $\kappa''$, $\kappa' - \kappa''$, or $\kappa' + \kappa''$. Since we only want to calculate the $N^2$ term of the intensity, we need *all* terms in the sum to contribute equally. This leads to the conditions

$$\kappa = \text{integer}$$
$$\text{or } \kappa - qa = \text{integer} \qquad \text{(II.1.23)}$$
$$\text{or } \kappa + qa = \text{integer.}$$

The corresponding intensities are

$$I = I_0 N^2 \left[ 1 - \left( \frac{\pi \kappa u_0}{a} \right)^2 \right] \qquad \text{for } \kappa = \text{integer} \qquad \text{(II.1.24)}$$

$$I = I_0 N^2 \frac{1}{2} \left( \frac{\pi \kappa u_0}{a} \right)^2 \qquad \text{for } \kappa \pm qa = \text{integer. (II.1.25)}$$

Note that the $\kappa = $ integer expression (Eq. II.1.24) is the expansion of

$$I = I_0 N^2 e^{-\frac{1}{2}(k u_0)^2} \qquad \text{(II.1.26)}$$

(remember $k = \frac{2\pi}{a} \kappa$). A more detailed inspection of the terms in the expansions given in Eqs. II.1.19 and II.1.20 reveal that Eq. II.1.26 is correct to every order of $k u_0$.

The diffraction pattern given by $I(k)$ will have strong peaks at integer values of $k$, with each strong peak having weaker ($I \propto u_0^2$) satellite peaks at $\pm qa$ away from it. As $u \rightarrow 0$, these side peaks will disappear. A static modulation of atomic position is observed in charge density wave systems like $TaS_3$, $NbSe_3$ or in the organic conductor TTF-TCNQ.[3]

## 1.12 Solution: Powder Diffraction of $hcp$ and $fcc$ Crystals

To satisfy the powder diffraction condition, the length of the reciprocal lattice vector must be equal to $G = K = 4\pi/\lambda \sin \theta$.

To calculate the position of the peaks for the $fcc$ structure we use the simple cubic unit cell with four atoms per cell. The reciprocal lattice is cubic, with lattice spacing of $G_0 = 2\pi/a_{cubic} = 1.77$ Å$^{-1}$, and the structure factor is nonzero only if the $(hkl)$ indices are all odd or all even.[4] The length of the reciprocal lattice vector is $G = G_0 \sqrt{h^2 + k^2 + l^2}$ (see Table II.1.4).

The $hcp$ lattice has two atoms per unit cell. The reciprocal lattice is constructed from a simple hexagonal lattice by assigning a zero structure factor to some of the points, resulting in alternating hexagonal and honeycomb arrays.[5] We will index the reciprocal lattice points in terms of $G = h a^* + k b^* + l c^*$, where the angle between $a^*$ and $b^*$ is 120°, and $c^*$ is perpendicular to $a^*$ and $b^*$. The lengths of the primitive vectors are calculated from Eq. I.1.1 as $a^* = b^* = 4\pi/\sqrt{3} \, a = 2.89$ Å$^{-1}$,

---

[3] For a more detailed discussion, see Grüner [16].
[4] see, for example, Kittel [2] p. 45.
[5] see, for example, Ashcroft and Mermin [1] p. 109.

**Table II.1.4.** Lengths of reciprocal lattice vectors for various $hkl$ in $\beta$-Co (*fcc* lattice).

| $hkl$ | $h^2 + k^2 + l^2$ | $G(\text{Å}^{-1})$ |
|-------|-------------------|---------------------|
| 111   | 3                 | 3.07                |
| 002   | 4                 | 3.54                |
| 220   | 8                 | 5.01                |
| 311   | 11                | 5.87                |
| 222   | 12                | 6.13                |
| 400   | 16                | 7.08                |
| 331   | 19                | 7.71                |
| 420   | 20                | 7.92                |

$c^* = 2\pi/c = \sqrt{3/8}\, 2\pi/a = 1.53\ \text{Å}^{-1}$. The lengths of reciprocal lattice vectors belonging to a few $(hkl)$ combinations are listed in Table II.1.5.

**Table II.1.5.** Lengths of reciprocal lattice vectors for various $hkl$ indices in $\alpha$-Co (which has an *hcp* lattice).

| $hkl$ | $G(\text{Å}^{-1})$ |
|-------|---------------------|
| 100   | 2.89                |
| 002   | 3.07                |
| 101   | 3.27                |
| 102   | 4.21                |
| 110   | 5.01                |
| 103   | 5.43                |
| 200   | 5.78                |
| 112   | 5.87                |
| 004   | 6.13                |

As discussed in Problem 1.6, the nearest-neighbor distance is the same for the two structures. This is why some of the X-ray diffraction peak positions coincide.

## 1.13 Solution: Momentum Resolution

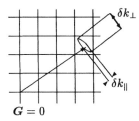

$G = 0$

**Fig. II.1.8.** Schematic representation of the reciprocal lattice and the incident neutron beam wavevector, $k_0$.

The reciprocal lattice of the cubic crystal is also cubic, with a reciprocal lattice spacing of $g = 2\pi/(3.55\ \text{Å}) = 1.76\ \text{Å}^{-1}$. The incident neutron beam is represented by a set of vectors, originating from $G = 0$, and pointing in various directions around $k_0$, as allowed by $\delta\phi$, with magnitudes set by $\delta\lambda$. A schematic drawing of this situation is shown in Figure II.1.8. The incident wavevector is $k_0 = 2\pi/\lambda = 2.51$ Å$^{-1}$, with $\delta k_\perp = k_0\delta\phi$ and $\delta k_\parallel = k_0\frac{\Delta\lambda}{\lambda}$. With the parameters of this problem, $\delta k_\perp/k_0 = \frac{40}{60}\frac{2\pi}{360} = 1.2 \times 10^{-2}$ and $\delta k_\parallel/k_0 = 8 \times 10^{-4}$. Clearly the "resolution ellipse" is long and narrow. The detector accepts neutrons from a similar ellipse, centered at $k$.

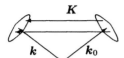

**Fig. II.1.9.** Neutron scattering with two possible choice of $K = k - k_0$.

The resolution can be represented by the diagram in Figure II.1.9. The best length resolution for $K$ can be attained by aligning $k$ and $k_0$ directly opposite to one another. This is called the "back-scattering" geometry. In this case,

$$\frac{\delta K_\perp}{K} = 2\frac{\delta k_0}{k_0} \quad \text{and} \quad \frac{\delta K_\parallel}{K} = 2\delta\phi . \qquad \text{(II.1.27)}$$

To be as close as possible to the back-scattering geometry we should look for a reciprocal lattice vector with a length close to but somewhat smaller than twice $k_0$, 5.02 Å$^{-1}$. The first several candidate reciprocal lattice vectors are given in Table II.1.6. The $(0, 2, 2)$ vector is the best choice.

| hkl | $G$ (Å$^{-1}$) |
|-----|------|
| 001 | 1.76 |
| 112 | 4.31 |
| 022 | 4.97 |
| 122 | 5.28 |

**Table II.1.6.** Lengths of Reciprocal Lattice Vectors of a Simple Cubic Material

For $G = 4.97$ Å$^{-1}$ and $k_0 = 2.51$ Å$^{-1}$, we obtain a scattering angle of $\theta = 82°$. The experimental scattering geometry is sketched in Figure II.1.10. The resolution is given by $\delta K/K = 1.6 \times 10^{-3} = \delta a/a$.

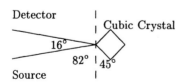

Detector

Cubic Crystal

Source

**Fig. II.1.10.** Sketch of the experimental scattering geometry. The diffraction from a powder sample is due to the properly oriented crystallites, as shown here.

## 1.14 Solution: Finite Size Effects

(a) Mathematically, the solution is similar to Solution 1.10, except here we have three dimensions. The amplitude of the scattered radiation is given by

$$E \propto \sum_{n=1}^{N_x \times N_y \times N_z} e^{iKR_n} S_K = \sum_{n=1}^{N_z} e^{iK_z R_{zn}} \times N_x \times N_y \ . \tag{II.1.28}$$

For simplicity we will take the structure factor $S_K = 1$. $R_{zn} = na$, where $a$ is the lattice spacing. Therefore the amplitude becomes

$$E \propto N_x N_y \sum_{n=1}^{N_z} \left(e^{iK_z a}\right)^n = N_x N_y \frac{e^{iK_z aN} - 1}{e^{iK_z a} - 1} e^{-iK_z a} \ , \tag{II.1.29}$$

where we have used the geometrical series result

$$\sum_{k=1}^{N} q^{k-1} = \frac{q^N - 1}{q - 1} \ . \tag{II.1.30}$$

Hence the intensity $I = | E |^2$ is given by

$$I \propto N_x^2 N_y^2 \frac{e^{iK_z aN} - 1}{e^{iK_z a} - 1} \frac{e^{-iK_z aN} - 1}{e^{-iK_z a} - 1} \tag{II.1.31}$$

$$= N_x^2 N_y^2 \frac{1 - \cos N_z K_z a}{1 - \cos K_z a} \ . \tag{II.1.32}$$

For $K_z - G_z = \delta \ll \frac{2\pi}{a}$ ($G_z$ is given by $\ell \frac{2\pi}{a}$, $\ell$ is an integer) we have

$$I \propto N_x^2 N_y^2 \frac{1 - \cos N_z a\delta}{\frac{1}{2} a^2 \delta^2} \ . \tag{II.1.33}$$

In the limit of $\delta \ll \frac{2\pi}{N_z a}$, the intensity becomes $I \propto N_x^2 N_y^2 N_z^2$. The constant $K_x$ cut of the function is shown in Figure II.1.11.

The integrated intensity is determined from

$$\int I \, dk_z \propto \int \frac{1 - \cos N_z a\delta}{\frac{1}{2} a^2 \delta^2} \, d\delta \tag{II.1.34}$$

$$= \int \frac{\sin^2 N_z a\delta/2}{\frac{1}{2} a^2 \delta^2} \, d\delta \tag{II.1.35}$$

$$= \frac{2}{a^2} \frac{N_z a\pi}{4} \ . \tag{II.1.36}$$

This result is only valid for $N_z \gg 1$. The integrated intensity is therefore proportional to $N_z$.

(b) Given a collection of $M$ crystallites, each having $N$ atoms, the amplitude of the scattered radiation will be

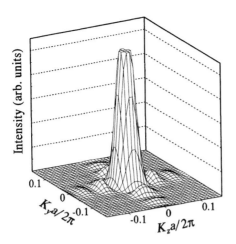

**Fig. II.1.11.** Intensity peak around $(1, 0, 0)$ for a cubic crystal of finite size in the $K_y, K_z$ plane.

$$E \propto \sum e^{iK(R_n + A_1)} + \sum e^{iK(R_n + A_2)} + \cdots + \sum e^{iK(R_n + A_M)} , \quad (\text{II}.1.37)$$

where $A_1 \ldots A_M$ are the positions of the crystallites and $R_n$ gives the positions of the atoms within each crystallite. We can rewrite this as

$$E \propto \left( \sum e^{iKR_n} \right) \left( e^{iKA_1} + e^{iKA_2} + \cdots + e^{iKA_M} \right) \quad (\text{II}.1.38)$$

$$= N \left( e^{i\phi_1} + e^{i\phi_2} + \cdots + e^{i\phi_M} \right) . \quad (\text{II}.1.39)$$

It now can be seen that this problem is equivalent to a two-dimensional random walk with steps of $e^{i\phi_n}$. The result is therefore $I = |E|^2 = N^2 M$. The width of the peak does not change. If the crystallites have different sizes, each one contributes a different amplitude in the random walk. The previous result will still hold where $N$ is the average crystal size.

## 1.15 Solution: Random Displacement

The amplitude of the scattered radiation is given by

$$E \propto \sum_n e^{iG \cdot (R_n + S_n)} = \sum_n e^{iG \cdot R_n} \cdot e^{iG \cdot S_n} . \quad (\text{II}.1.40)$$

If $G$ is a reciprocal lattice vector, then the the first term is always equal to 1, and

$$E \propto e^{iG \cdot S_n} = \sum_n e^{i\phi_n} , \quad (\text{II}.1.41)$$

where $\phi_n = G \cdot S_n = G S_n$ is a random phase with $S_n$ being the component of the random displacement parallel to $G$. The magnitude of the displacements is characterized by $S_0 = \sqrt{\overline{S_n^2}}$, where $\overline{S_n^2} = (\sum S_n^2)/N$.

First we look at the case of small $G$. For example, for $G = 2\pi/a$ we have $\phi_n \ll 2\pi$, since $S_n \ll a$. Therefore $e^{i\phi_n} = 1 + i\phi_n - \frac{1}{2}\phi_n^2$. One obtains, by adding up the electric field contributions,

$$E \propto N \left( 1 + i\overline{\phi} - \frac{1}{2}\overline{\phi^2} \right) , \tag{II.1.42}$$

where the average phase is $\overline{\phi} = (\sum \phi_n)/N = 0$ and $\overline{\phi^2} = (\sum \phi_n^2)/N = G^2 S_0^2$. The intensity is

$$I = \mid E \mid^2 \propto N^2 \left( 1 - \overline{\phi^2} \right) = N^2 \left( 1 - G^2 \overline{S_n^2} \right) = N^2 (1 - G^2 S_0^2) . \tag{II.1.43}$$

We note that the peak height of the intensity scales with $N^2$; the width of the peaks are of the order of $1/N$, as discussed in Problem 1.14.

For large G, the peak intensity will vanish. For $G > 2\pi/S_0$ the summation of the electric field vectors is equivalent to a two-dimensional random walk problem, resulting in

$$E \propto \sqrt{N} \quad \longrightarrow \quad I \propto N . \tag{II.1.44}$$

The scattering peak height will diminish relative to the low $G$ value by a factor of $1/N$.

To obtain a general solution we first replace the sum in Eq. II.1.41 by an integral. Let us assume that the probability of finding $S_n$ between $S$ and $S + dS$ is given by $f(S)dS$. The distribution function is normalized such that $\int f(S)dS = 1$

The field is then

$$E \propto N \int_{-\infty}^{\infty} f(S)e^{iGS} \, dS . \tag{II.1.45}$$

The result depends on the actual distribution function $f(S)$. If each individual atomic displacement is produced by many random events, then the central limit distribution theorem guarantees that $f(S) = \frac{1}{\sqrt{2\pi}S_0}e^{-S^2/2S_0^2}$, a Gaussian distribution. Inserting this to the integral we get

$$I \propto E^2 \propto e^{-G^2 S^2} . \tag{II.1.46}$$

The intensities of the peaks at large $G$ drop rather fast. This would seem to contradict Solution 1.17, where it was shown that the total integrated scattering intensity should not change if the atomic positions change. However, in the present solution we have only considered the intensity of the Bragg peaks. The "missing" intensity will appear as a broad, incoherent scattering background.

## 1.16 Hint: Vacancies

The vacancies will cause a decrease in the Bragg peak intensities, and will also cause a weakly and smoothly angular-dependent background scattering.

## 1.17 Hint: Integrated Scattering Intensity

The intensity is given by

$$I = (A_0 \sum_n e^{iKR_n})(A_0^* \sum_{n'} e^{-iKR'_n}) =\mid A \mid^2 \sum_{n,n'} e^{iK(R_n - R_{n'})} \ . \qquad (\text{II.1.47})$$

Integration over $K$ leads to a delta function in $R_n - R_{n'}$, and the sum over $n$ and $n'$ becomes independent of $R$.

Note that this theorem is only valid if the scattering is perfectly elastic.

# 2 Interatomic Forces, Lattice Vibrations

## 2.1 Solution: Madelung Constant

The total Coulomb energy of a solid is given by

$$U = \frac{1}{2} \sum_{i,j} \sum_{\mu,\nu} \frac{q_{i\mu} q_{j\nu}}{r_{i\mu,j\nu}}, \tag{II.2.1}$$

where $i$ and $j$ sum over various unit cells, and $\mu$ and $\nu$ sum over the atoms within one unit cell. If the total number of unit cells is $N$, then the sum can be simplified to

$$U = \frac{1}{2} N \sum_{j} \sum_{\mu,\nu} \frac{q_{1\mu} q_{j\nu}}{r_{1\mu,j\nu}}. \tag{II.2.2}$$

The structure of $Rb_3 C_{60}$ is face centered cubic (see Problem 1.5). For a uniform charge distribution over the fullerenes, the energy calculation can be performed by assuming that the charge is concentrated in the center of the fulleride ion. If the $C_{60}^{3-}$ ions are positioned at the corners and face centers of a cube, then the alkali metal ions will be at the midpoints of the cube edges, and also at one quarter of the body diagonals. In terms of coordinates in a cube of size $a = 2$, the $C_{60}$ ions will be at the $(0,0,0)$, $(0,0,2)$, $(1,1,0)$ .... positions. Alkali ions will be at the so called octahedral positions $[(1,0,0),$ $(0,1,0)$ ....], and at the tetrahedral positions at $[(1/2,1/2,1/2), (1/2,-1/2,1/2)$ ....].

The definition of the Madelung constant given in Eq. I.2.5, (and in many introductory solid state textbooks) implicitly assumes that there are equal number of positive and negative ions in the system. In that case $M = u/u_0$, where $u$ is the potential energy of a single ion in the solid, and $u_0 = e^2/d$ is the binding potential of a single positive–negative ion pair, positioned at a distance $d$, the nearest neighbor distance between the two ions in the solid. In our case a similar comparison is not very meaningful. (One $C_{60}^{3-}$ ion and one $Rb^+$ ion does not constitute a neutral pair; if we begin to place the other two Rb ions around the fullerene, the result will depend on the relative positions of the three positive charges; we have many equally appealing choices for these positions, and they will lead to different reference energies $u_0$.)

In a somewhat arbitrary manner we will define the Madelung constant here as the coefficient $M$ in

$$U = \frac{1}{2} N \sum_j \sum_{\mu,\nu} \frac{q_{1\mu} q_{j\nu}}{r_{1\mu,j\nu}} = \frac{e^2}{d} \frac{1}{2} N \sum_j \sum_{\mu,\nu} \frac{Q_{i\mu} Q_{j\nu}}{p_{1\mu,j\nu}}$$

$$= MN \frac{e^2}{d}, \tag{II.2.3}$$

where $Q = \pm 1$ represents the charge, and $p$ is in units of $d$, a characteristic length in the crystal. For the $fcc$ lattice of $Rb_3 C_{60}$ we will chose $d = a/2$, where $a$ is the cubic lattice spacing. With this choice, $d$ is the distance between the center of the $C_{60}$ and the octahedral Rb ion.

The convergence of the Madelung sum is best established by physical arguments. First, a real solid is charge-neutral, and therefore we need to have equal number of positive and negative terms in the sum. However, this is not enough! We also need to make sure that we do not have uncompensated surface charges causing internal electric fields. In a computer program we can inadvertently put pairs of positive and negative charges together in such a manner that the result is a huge, charged capacitor. The energy, of course, would not converge, but it would be unphysically proportional to the volume.

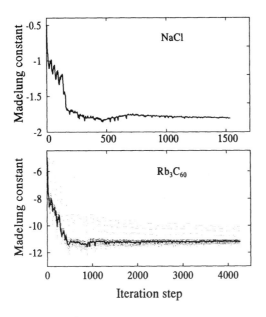

Fig. II.2.1. The convergence of the Madelung energy calculation for two solids of $fcc$ structure. The solid line represents the sum when there is zero total charge; the dots show the scatter in the result as the sum proceeds. The iteration steps progress from the closest to the farthest unit cells.

Figure II.2.1 shows the results of our computer calculations for NaCl and for $Rb_3 C_{60}$. For NaCl the "crystal" was built of a triplet of $-1/2 - +1 - -1/2$ charges, spaced at a distance of 1. This triplet was then repeated at the points $(i, j, k)$ where $i$, $j$, and $k$ are integers. The $fcc$ structure was implemented by considering only the points where $i + j + k$ is an even number. During this procedure, the fractional charges in the unit cell will add up to integer charges at their correct positions.

The value of the sum fluctuates considerably, but when the sum has covered a "charge neutral" amount of the crystal, the convergence is fast.[1] The crucial step was using a building block which did not have dipole moment. Our calculated Madelung constant for NaCl is 1.75 in agreement with the known result (Ashcroft and Mermin [1] p. 405).

Similarly, the fulleride crystal was constructed from a $q = +3$ charge at $(0,0,0)$, $q = -1/2$ charges at the $(-1/2,0,0)$ and $(1/2,0,0)$ points, and $q = -1$ charges at $(1/4,1/4,1/4)$ and $(-1/4,-1/4-1/4)$. Again, this is not a very symmetric arrangement, but it has zero dipole moment and does the job.

The resultant Madelung constant of $Rb_3C_{60}$ is 11.04.

## 2.2 Solution: NaCl Bulk Modulus

Using the interaction potential given in the problem, the total energy of the system is written as

$$U = N \left( -\frac{e^2}{r} \sum \frac{\pm 1}{\rho_{ij}} + \frac{\alpha}{r^n} \sum \frac{1}{\rho_{ij}^n} \right) , \qquad \text{(II.2.4)}$$

where $\rho_{ij} = r_{ij}/r$ and $r = 2.10$ Å is the distance between Na and Cl first-nearest-neighbors. The first sum is for alternating charged ions ($Na^+$ and $Cl^-$); it yields, by definition, the Madelung constant, $M = 1.748$.[2] The second sum results in $\alpha \sum \frac{1}{\rho_{ij}^n} = C$.

Pressure is given by $P = -\frac{\partial U}{\partial V}$. The volume is $V = 2Nr^3$. Therefore $dV = 6Nr^2 dr$, and we can solve for the pressure,

$$P = -\frac{1}{6Nr^2} \frac{\partial}{\partial r} N \left( -Me^2 \frac{1}{r} + \frac{C}{r^n} \right) \qquad \text{(II.2.5)}$$

$$= -\frac{1}{6r^2} \left( ME^2 \frac{1}{r} - \frac{nC}{r^{n+1}} \right) . \qquad \text{(II.2.6)}$$

The equilibrium lattice parameters are determined at $P = 0$ ($P = 1$ atm is essentially zero). Note that this is equivalent to finding a minimum in $U(r)$:

$$0 = \frac{Me^2}{r_0^2} - \frac{nC}{r_0^{n+1}} . \qquad \text{(II.2.7)}$$

---

[1] The distribution of charges over a cubic unit cell, as described by Ashcroft and Mermin [1] p. 404, requires more attention when creating the unit cell. On the other hand, these unit cells form a simple cubic lattice.

[2] See Ashcroft and Mermin [1] p. 405.

The compressibility is determined from

$$B \;=\; -V\frac{\partial P}{\partial V} = 2Nr^3\frac{1}{6Nr^2}\frac{\partial}{\partial r}\frac{1}{6r^2}\left(ME^2\frac{1}{r} - \frac{nC}{r^n}\right) \tag{II.2.8}$$

$$=\; \frac{1}{18}r\frac{\partial}{\partial r}\left(Me^2\frac{1}{r^4} - \frac{nC}{r^{n+3}}\right) \tag{II.2.9}$$

$$=\; \frac{1}{18}\left(\frac{n(n+3)}{r_0^{n+3}}C - Me^2\frac{4}{r_0^4}\right) . \tag{II.2.10}$$

Using Eqs. II.2.7 and II.2.10, we have two equations from which to solve for two unknowns, $C$ and $n$. We can rewrite Eq. II.2.7 as

$$\frac{C}{r_0^{n+3}} = Me^2\frac{1}{nr_0^4} , \tag{II.2.11}$$

and Eq. II.2.10 becomes

$$B \;=\; \frac{Me^2}{18}\left(\frac{n(n+3)}{n} - 4\right)\frac{1}{r_0^4} \tag{II.2.12}$$

$$=\; \frac{1}{18}Me^2\frac{1}{r_0^3}(n-1) . \tag{II.2.13}$$

We have separated out the terms because we know that $\phi_{\text{Coulomb}} = \frac{e^2 A}{4\pi\varepsilon_0 r_0}$, and therefore we can obtain

$$n = 1 + \frac{18r_0^3 B}{\phi_{\text{Coulomb}}} . \tag{II.2.14}$$

Using the values given in the problem, $\phi_{\text{Coulomb}} = 8.53$ eV and

$$n = 1 + \frac{18 \cdot (2.82 \times 10^{-8}\ \text{cm})^3 \cdot 2.4 \times 10^{11}\ \frac{\text{dyn}}{\text{cm}^2}}{8.53 \cdot 1.6 \times 10^{-12}} = 8.1 . \tag{II.2.15}$$

From Eq. II.2.7 we can see that

$$\phi_{\text{Coulomb}} = \frac{nC}{r_0^n} \;\longrightarrow\; C = \frac{\phi_{\text{Coulomb}}r_0^n}{n} , \tag{II.2.16}$$

so $C = 1.05$ eV$\cdot(4436\ \text{Å})^{8.1} = 4660$ eVÅ$^{8.1}$. To obtain $\alpha$ from $C$, we have to calculate $\sum\frac{1}{\rho_{ij}^n}$. There are six first neighbors which have $\rho_{ij} = 1$, 12 second nearest-neighbors ($\rho_{ij} = \sqrt{2}$), eight third nearest-neighbors ($\rho_{ij} = \sqrt{3}$), …:

$$\sum\frac{1}{\rho_{ij}^n} = 6 + 12\frac{1}{2^{4.05}} + 8\frac{1}{3^{4.05}} + \cdots = 6.81 . \tag{II.2.17}$$

Therefore,

$$\alpha = \frac{C}{6.81} = 684\ \text{eVÅ}^{8.1} . \tag{II.2.18}$$

## 2.3 Hint: Madelung with Screened Potential

Screening in a linear chain dramatically improves the convergence of the sum in the Madelung energy calculation. This is the best algorithm to investigate the $\alpha \to 0$ limit. Calculate the Madelung sum with $\alpha = 0.5 \times 1/d$ (where $d$ is the nearest-neighbor distance) for $N$ terms. The result will oscillate as a function of $N$, but the amplitude of the oscillations will decrease rapidly as $N$ increases. Continue the calculation until the oscillation amplitude decreases below a predetermined value, let's say $0.01e^2/d$. Repeat the calculation with several, subsequently smaller values of $\alpha$, until the convergence becomes slow (what "slow" is depends on your patience and computing power). The result will be $M = M(\alpha)$. Plot this function, and extrapolate $M$ to $\alpha = 0$ to obtain the "unscreened" Madelung constant value.

## 2.4 Solution: Triple-axis Spectrometer

The energy resolution is given by

$$\delta E = E_0 \frac{2\delta\lambda_0}{\lambda_0} + E' \frac{2\delta\lambda}{\lambda} , \qquad (\text{II}.2.19)$$

where $E_0 = \hbar^2 k_0^2/(2m) = 2.07 k_0^2$ meV $= 13.06$ meV, and the energy of the scattered neutrons is $E' = E_0 - 5$ meV $= 8.06$ meV. The numerical value of the resolution is therefore $\delta E = 2 \times 13.06 \times 8 \times 10^{-3} = 3.7 \times 10^{-2}$ meV.

## 2.5 Solution: Phonons in Silicon

The sound velocity is the slope of the appropriate acoustic mode at $k = 0$. For the calculation, one has to take into account that the horizontal scale of the spectrum is often given in units of the reciprocal lattice vector. It is also important to clarify if the vertical scale of the figure is in units of frequency $f$, or angular frequency $\omega$.

Let us take $\xi = 0.2$ along the (100) direction. The frequency of the transverse mode is $f = 2 \times 10^{12}$ sec$^{-1}$. Since $q = \xi\frac{2\pi}{a}$, and the lattice spacing is $a = 5.43$ Å, we have $q = 0.23$ Å$^{-1}$. The slope is

$$c = \frac{d\omega}{dq} \approx \frac{2\pi \times 2 \times 10^{12}}{0.23} = 5400 \text{ m/sec} \qquad (\text{II}.2.20)$$

For the $(\xi\xi 0)$ and $(\xi\xi\xi)$ directions the calculation is similar.

The measured sound velocities are tabulated in Landolt–Börnstein [21] Vol. 17, p. 63. Note that the sound velocity is not isotropic.

## 2.6 Solution: Linear Array of Emitters, Phonons

Making the modulation time-dependent, we calculate the time-dependent intensity at the detector,

$$I(t) = I_0 \cos(\omega t) \sum_{n,m} e^{i2\pi(n-m)\kappa} \cdot e^{i2\pi\kappa(u_n(t)-u_m(t))/a} \ , \qquad (\text{II.2.21})$$

where $\omega$ is the frequency being emitted by the sources. We now expand as in Problem 1.11, inserting

$$u_n(t) = \frac{u_0}{2} \left( e^{i(nqa-\Omega t)} + e^{-i(nqa-\Omega t)} \right). \qquad (\text{II.2.22})$$

All the considerations that went into the expansion before are still valid, but whenever a $qa$ appears in the equations, we have to include an $e^{i\Omega t}$ factor.

This leads to solutions analogous to Eq. II.1.26:

$$I = I_0 N^2 e^{-\frac{1}{2}(ku_0)^2} \qquad \text{at } \omega_{\text{detector}} = \omega \qquad (\text{II.2.23})$$

$$I = I_0 N^2 \frac{1}{4} \left( 1 - e^{-\frac{1}{2}(ku_0)^2} \right) \qquad \text{at } \omega_{\text{detector}} = \omega \pm \Omega. \quad (\text{II.2.24})$$

The frequencies and wavenumbers where radiation will be detected is sketched in Figure II.2.2.

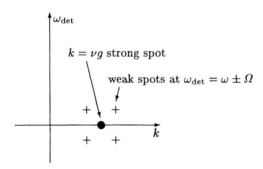

**Fig. II.2.2.** Points in frequency-wavenumber space where radiation will be detected for a time-dependent modulation of a linear array of radiation sources.

## 2.7 Hint: Long Range Interaction

In one dimension the solution of Eq. I.2.10 is trivial, $D(k) = \lambda(k)$. The phonon frequencies are $w(k) = \sqrt{D(k)/M}$, and for a linear dispersion, $\omega = ck$, we obtain $D(k) \sim k^2$. The spring constants are obtained from the equation $D_{i,j} \sim \int_{-\pi/a}^{\pi/a} e^{ikR} k^2 dk$.

## 2.8 Solution: Mass Defect

If the displacement of the $n$th mass is $s_n$, the equations of motion for the masses are

$$m\ddot{s}_{-1} = \kappa(s_{-2} - 2s_{-1} + s_0) \tag{II.2.25}$$

$$M\ddot{s}_0 = \kappa(s_{-1} - 2s_0 + s_1) \tag{II.2.26}$$

$$m\ddot{s}_1 = \kappa(s_0 - 2s_1 + s_2) \tag{II.2.27}$$

or more generally

$$m\ddot{s}_n = \kappa(s_{n-1} - 2s_n + s_{n+1}) \text{ for } n \neq 0 . \tag{II.2.28}$$

Applying the ansatz

$$s_n = s_0 e^{k(w)|n| - i\omega t}, \tag{II.2.29}$$

we obtain

$$s_n = s_0 e^{-kn - i\omega t} \qquad \text{for } n > 0 \tag{II.2.30}$$

$$s_n = s_0 e^{-i\omega t} \qquad \text{for } n = 0 \tag{II.2.31}$$

$$s_n = s_0 e^{+kn - i\omega t} \qquad \text{for } n < 0. \tag{II.2.32}$$

Inserting this into the equations of motion, we obtain

$$-m\omega^2 e^{-kn} = \kappa e^{-kn}(e^k - 2 + e^{-k}) \qquad \text{for } n > 0 \tag{II.2.33}$$

$$-M\omega^2 = \kappa(e^{-k} - 2 + e^{-k}) \qquad \text{for } n = 0 \tag{II.2.34}$$

$$-m\omega^2 e^{kn} = \kappa e^{kn}(e k - 2 + e^{-k}) \qquad \text{for } n < 0. \tag{II.2.35}$$

This reduces to two equations,

$$-M\omega^2 = 2\kappa(1 - e^{-k}) \tag{II.2.36}$$

$$-m\omega^2 = \kappa(2 - e^k - e^{-k}) , \tag{II.2.37}$$

whose solutions are

$$\omega = 2\sqrt{\frac{\kappa}{m}}\sqrt{\frac{m/M}{2 - M/m}} \tag{II.2.38}$$

and $\quad k = \ln\left(1 - \dfrac{2m}{M}\right) .$ $\tag{II.2.39}$

*Discussion.*

- For $M > 2m$ $\omega$ is imaginary and $k$ is negative and real. This means that our trial function did not work: The solution decays in time and explodes in space.

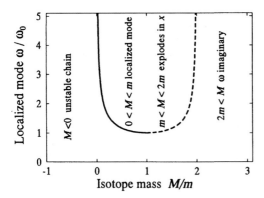

**Fig. II.2.3.** Localized mode frequency as a function of the isotope mass. The oscillatory solution is nondivergent only for $0 < M < m$.

- For $M < 2m$ the frequency is real, and the parameter $k$ is complex. This is an oscillatory solution, but we have to decide if the position dependence of $s_n$ is acceptable. Using the property $\ln(-A) = i\pi + \ln A$ we obtain

$$k = i\pi + \ln\left(\frac{2m}{M} - 1\right). \qquad (\text{II.2.40})$$

- For $m < M < 2m$ Re$(k)$ is negative and $s_n$ diverges for large $n$. The trial function does not work.
- For $M < m$ Re$(k)$ is positive. This is when the trial function is applicable. The solution describes oscillating particles. The amplitude decays with the distance from the isotope.

In summary, our ansatz only works for $0 < M < m$. (For negative mass $M$ the chain would be unstable.) The time dependence of the solution is oscillatory; the spatial dependence is a decaying wave. Note that the frequency of the oscillation is $\omega \geq 2\sqrt{\frac{\kappa}{m}}$, whereas for the defect-free chain $(M = m)$, $\omega = 2\sqrt{\frac{\kappa}{m}}$. The frequency of the localized mode is *above* the highest phonon mode of the pure lattice. This is expected, since a lighter mass usually means a higher oscillation frequency. These results are summarized in Figure II.2.3

To solve the problem for other values of $M$, a different ansatz must be chosen. Ziman [3] (pp. 71–76) and Harrison [5] (pp. 381–389) discuss the general solution in detail.

## 2.9 Solution: Debye Frequency

The Debye frequency is determined by two conditions:

- the total number of modes
- the sound velocity.

From the total number of modes we obtain the Debye wavenumber, $k_D$,

$$\frac{4\pi}{3}k_D^3\left(\frac{L}{2\pi}\right)^3 = N \implies k_D = (6\pi^2 n)^{1/3}, \qquad (\text{II}.2.41)$$

where $n$ is the number of modes per unit volume ($n = \frac{N}{L^3}$).

The sound velocity converts the Debye wavenumber to the Debye frequency,

$$\omega_D = ck_D = c(6\pi^2 n)^{1/3}. \qquad (\text{II}.2.42)$$

Although $\omega^\star$ and $\omega_D$ are typically of the same order of magnitude, there is no firm relationship between the two quantities, and $\omega^\star = \omega_D$ happens only by accident.

## 2.10 Solution: Vibrations of a Square Lattice

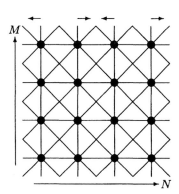

Fig. II.2.4. Sketch of the longitudinal mode (denoted by the arrows on top of the figure) in a monoatomic two-dimensional square lattice.

We draw the motion of the atoms in this square lattice for the longitudinal modes in Figure II.2.4. We denote the displacements by $u$ and express the kinetic and potential energies of the system in terms of first and second derivatives of $u$, $\dot{u}$, and $\ddot{u}$, respectively. The kinetic energy comes from $N \times M$ atoms, moving with the same velocity, $|\dot{u}|$;

$$E_K = \frac{1}{2}m\dot{u}^2 NM. \qquad (\text{II}.2.43)$$

The potential energy comes from $N \times M$ springs with spring constant $k_1$ compressed by $2u$, and from $2N \times M$ springs with spring constant $k_2$ compressed by $2u/\sqrt{2}$:

$$E_P = \frac{1}{2}(4k_1 + 4k_2)u^2 NM. \qquad (\text{II}.2.44)$$

The total energy is therefore

$$E = E_K + E_P = \frac{1}{2}NM\left[m\dot{u}^2 + (4k_1 + 4k_2)u^2\right]. \qquad (\text{II}.2.45)$$

We can solve for the frequency of this longitudinal mode just by analogy to a simple ball and spring model ($E = \frac{1}{2}M\dot{x}^2 + \frac{1}{2}Kx^2$ has a frequency of $\omega = \sqrt{K/M}$). Therefore $\omega_L$ is

$$\omega_L = 2\sqrt{\frac{k_1 + k_2}{m}} \; . \tag{II.2.46}$$

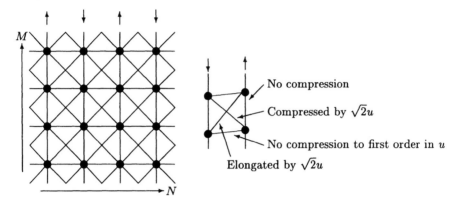

**Fig. II.2.5.** Left figure is a sketch of the transverse mode (denoted by the arrows on top of the figure) in a monoatomic two-dimensional square lattice. Right figure shows the spring compressions associated with this transverse motion.

For the transverse mode, the displacements and corresponding spring compressions are shown in Figure II.2.5. The kinetic energy is again given by

$$E_K = \frac{1}{2}m\dot{u}NM. \tag{II.2.47}$$

The potential energy is given by (refer to the right part of Figure II.2.5 to see the spring compressions)

$$E_P = \frac{1}{2}k_2(\sqrt{2}u)^2 2NM. \tag{II.2.48}$$

We then solve for the frequency of this mode just as before and obtain

$$\omega_T = 2\sqrt{\frac{k_2}{m}} \; . \tag{II.2.49}$$

## 2.11 Solution: Grüneisen Parameter

The Grüneisen parameter is defined as

$$\gamma = -\frac{L}{\omega}\frac{d\omega}{dL} \ . \tag{II.2.50}$$

In a one-dimensional chain the frequency scale of the phonon dispersion curve $\omega(k)$ is determined by a single frequency parameter $\omega_0 = \sqrt{\kappa/m}$. Our goal is to determine how this frequency depends on the length of the chain. To induce a small change in the length, from $L$ to $L'$, we apply a force $\Delta F$. Then we will calculate the second derivative of the potential around the new equilibrium position, $\kappa' = d^2U/dx^2$. We use this new restoring force to calculate the frequency.

If a force is applied, the potential becomes

$$U = U_0 + \frac{1}{2}\kappa x^2 + \lambda x^3 + \Delta F x \ . \tag{II.2.51}$$

The new equilibrium position can be found from $dU/dx = 0$:

$$\kappa \Delta x + 3\lambda(\Delta x)^2 + F = 0 \ . \tag{II.2.52}$$

For a small force, the solution is $\Delta x = -\Delta F/\kappa$. The new lattice spacing is $a' = a + \Delta x_0$; the change in the length of the system is $\Delta L = N x_0$.

The restoring force is given by

$$\kappa' = d^2U/dx^2 = \kappa + 6\lambda\Delta x \ . \tag{II.2.53}$$

and the new frequency is

$$\omega_0' = \sqrt{\kappa'}/m = \omega_0 + \frac{1}{2}\frac{6\lambda}{\sqrt{\kappa m}}\Delta x \ . \tag{II.2.54}$$

Therefore we obtain the result

$$\gamma = -\frac{L}{\omega_0}\frac{\Delta\omega_0}{\Delta L} = -\frac{Na}{\omega_0 N\Delta x}\frac{3\lambda}{\sqrt{\kappa m}}\Delta x = -3\lambda a/\kappa \ . \tag{II.2.55}$$

The parameters $\lambda$ and $\kappa$ are coefficients in a Taylor series for the potential $U(x)$. In dimensionless units, for nonsingular functions, these coefficients are of the order of unity. Therefore $\gamma$ is of the order of unity. For typical interaction potentials (see, for example, Eqs. I.2.4 or I.2.2) $\lambda < 0$ and $\gamma > 0$.

## 2.12 Solution: Diatomic Chain

(a) The sound velocity is given by

$$c = \frac{\omega}{q} \qquad \text{for } q \to 0 . \tag{II.2.56}$$

Take the acoustic branch of the given phonon dispersion relation (the $-$ sign choice in Eq. I.2.16). As $q \to 0$, $\sin^2(\frac{qa}{2}) \to (\frac{qa}{2})^2$ and

$$\sqrt{a+x} = \sqrt{a}\sqrt{1 + \frac{x}{a}} \to \sqrt{a}\left(1 + \frac{1}{2}\frac{x}{a}\right) . \tag{II.2.57}$$

The phonon dispersion in this limit is then written

$$\omega^2 = f\left(\frac{1}{M_1} + \frac{1}{M_2}\right) - f\left(\frac{1}{M_1} + \frac{1}{M_2}\right)\left[1 - \frac{1}{2}\frac{\frac{q^2 a^2}{4 M_1 M_2}}{\left(\frac{1}{M_1} + \frac{1}{M_2}\right)^2}\right] \tag{II.2.58}$$

$$= f\frac{1}{8}\frac{q^2 a^2}{M_1 + M_2} . \tag{II.2.59}$$

Therefore the sound velocity is

$$c = \frac{\omega}{q} = \frac{qa}{2}\sqrt{\frac{f}{2(M_1 + M_2)}}\frac{1}{q} = \frac{a}{2}\sqrt{\frac{f}{2(M_1 + M_2)}} . \tag{II.2.60}$$

(b) For $M_1 = M_2$, we have a dispersion relation given by

$$\omega^2 = f\frac{2}{M} \pm f\sqrt{\left(\frac{2}{M}\right)^2 - \frac{4}{M^2}\sin^2\left(\frac{qa}{2}\right)} \tag{II.2.61}$$

$$= f\frac{2}{M}\left[1 \pm \sqrt{1 - \sin^2\left(\frac{qa}{2}\right)}\right] \tag{II.2.62}$$

$$= f\frac{2}{M}\left[1 \pm \cos\left(\frac{qa}{2}\right)\right] , \tag{II.2.63}$$

where the last equation is only valid for $-\frac{\pi}{2} < q < \frac{\pi}{2}$. We can further simplify this relation by using the trigonometric properties

$$\sin^2\left(\frac{x}{2}\right) = \frac{1}{2}(1 - \cos x) \tag{II.2.64}$$

$$\cos^2\left(\frac{x}{2}\right) = \frac{1}{2}(1 + \cos x) \tag{II.2.65}$$

to obtain

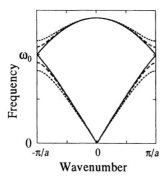 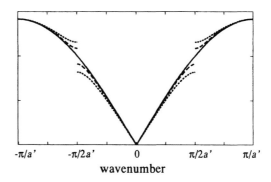

**Fig. II.2.6.** Phonon dispersion relation for a diatomic chain with various values of $\alpha = M_1/M_2$. For $M_1 = M_2$ $a' = a/2$ becomes the new lattice spacing and the dispersion is equivalent to the that of a monoatomic chain (solid line).

$$\omega_1^2 = \frac{4f}{M} \sin^2 \left( \frac{qa}{4} \right) \tag{II.2.66}$$

$$\omega_2^2 = \frac{4f}{M} \cos^2 \left( \frac{qa}{4} \right). \tag{II.2.67}$$

The lattice spacing is now $a' = a/2$ since the new lattice has only one atom per unit cell. Therefore, in terms of the new lattice spacing, the dispersions are

$$\omega_1 = \sqrt{\frac{4f}{M}} \sin \left( \frac{qa'}{2} \right) \tag{II.2.68}$$

$$\omega_2 = \sqrt{\frac{4f}{M}} \cos \left( \frac{qa'}{2} \right), \tag{II.2.69}$$

where this is valid for $-\frac{\pi}{2a'} < q < \frac{\pi}{2a'}$. Having the function over $-\frac{\pi}{a'} < q < \frac{\pi}{a'}$ would preferable; this can be achieved by shifting the upper branch of the dispersion by $\frac{\pi}{2}$, as demonstrated graphically in Figure II.2.6. This resultant dispersion is equivalent to the monoatomic chain result,

$$\omega = \sqrt{\frac{4f}{M}} \sin \left( \frac{qa'}{2} \right). \tag{II.2.70}$$

## 2.13 Hint: Damped Oscillation

The damping term will add an imaginary part to the algebraic equation derived from the equation of motion: $\omega^2 = (2\kappa/m)(1 - \cos ka) + i\omega \Gamma/m$. For high frequency modes ($k \approx \pi/a$) the mode frequency will be down-shifted by $\Gamma^2/8m\omega$ and the lifetime will be $2m/\Gamma$. The low frequency modes ($k \approx 0$) will cease to exist; no oscillatory component will remain in the solution.

## 2.14 Solution: Two-Dimensional Debye

The definition of Debye frequency is

$$V \int_0^{\omega_D} g(\omega)d\omega = N \, , \tag{II.2.71}$$

where $N$ is the total number of phonon modes and $g(E)$ is the density of modes with $\omega = ck$ dispersion.

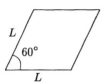

**Fig. II.2.7.** A cell of the hexagonal lattice.

$L$

$60°$

$L$

Let us take a lattice cell as shown in Figure II.2.7. The "volume" of this cell is $V = L^2 \cos 60°$; the total number of states is $(L/a)^2$. The "grid" in reciprocal space to be summed over is shown in Figure II.2.8.

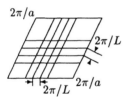

$2\pi/a$

$2\pi/L$

$2\pi/L$    $2\pi/a$

**Fig. II.2.8.** "Grid" in the reciprocal lattice.

We can obtain the density of states by summing over the reciprocal space grid; replacing the $\sum_k$ with an integral over d$k$, we obtain

$$\int \frac{dk}{\Delta^2 k} = \frac{1}{\left(\frac{2\pi}{L}\right)^2 \cos 60°} \int 2\pi k \ dk \, . \tag{II.2.72}$$

Then using the phonon dispersion, $\omega = ck$ and $d\omega = cdk$, we obtain

$$V g(\omega)d\omega = \frac{L^2}{\cos 60°} \frac{1}{2\pi} \frac{1}{c^2} \omega \ d\omega. \tag{II.2.73}$$

Therefore,

$$g(\omega) = \frac{1}{\cos 60°} \frac{1}{2\pi} \frac{\omega}{c^2} \, . \tag{II.2.74}$$

Substituting this density of modes into the integral in Eq. II.2.71, we find

$$L^2 \cos 60° \int_0^{\omega_D} \frac{1}{\cos 60°} \frac{1}{2\pi} \frac{\omega}{c^2} \ d\omega = \frac{L^2}{a^2} \, . \tag{II.2.75}$$

Another definition of Debye frequency is based on the density of states, as discussed in Chapter 4:

$$V \int_0^{\omega_D} g(\omega)d\omega = N \ , \tag{II.2.74}$$

where $N$ is the total number of phonon states and $g(\omega)$ has to be determined assuming $\omega = ck$ dispersion.

To obtain $g(\omega)$, we follow Eq. I.4.1, and we replace the $\sum_k$ with an integral over d$k$. We obtain

$$\int \frac{dk}{v^*/N} F = \frac{Na^2 \sin 60°}{4\pi^2} \int 2\pi k \ F \ dk \ = V \int g(\omega)F(\omega)d\omega. \tag{II.2.75}$$

Here $F$ is an arbitrary function of the frequency $\omega$ and the last term defines $g(\omega)$. The angular integration over the direction of the $k$ vectors yielded a factor of $2\pi$ since the phonon dispersion was assumed to be isotropic. Using $\omega = ck$ and $d\omega = cdk$, we obtain

$$V \int Fg(\omega)d\omega = \int F \frac{Na^2 \sin 60°}{2\pi} \frac{\omega}{c^2} \ d\omega \ , \tag{II.2.76}$$

therefore

$$g(\omega) = \frac{N}{V} \frac{\sqrt{3}}{8\pi} \frac{a^2 \omega}{c^2} \ . \tag{II.2.77}$$

Substituting this density of modes into the integral in Eq. II.2.74, we find

$$\frac{1}{2}\omega_D^2 = \frac{4\pi}{\sqrt{3}} \frac{c^2}{a^2} \ , \tag{II.2.78}$$

and therefore the Debye frequency is

$$\omega_D = 3.81 \ \frac{c}{a} \ . \tag{II.2.79}$$

## 2.15 Solution: Soft Optical Phonons

Assume that site $m$ has a polarization given by

$$p_m = p^* e^{ikam} e^{-i\omega t} \ . \tag{II.2.80}$$

The electric field at site $n$ due to site $m$ is

$$E_{nm} = \frac{2p_m}{d_{nm}^3} \ , \tag{II.2.81}$$

where $d_{nm} =| a(n - m) |$, the distance between $m$ and $n$. The total field at site $n$ due to all other sites is therefore

$$
\begin{aligned}
E_n &= \sum_{m \neq n} \frac{2p_m}{d_{nm}^3} \\
&= \frac{2p^*}{a^3} \sum_{m \neq n} \frac{e^{ikam} e^{-i\omega t}}{|n-m|^3} \\
&= e^{ikan} e^{-i\omega t} \frac{2p^*}{a^3} \sum_{m \neq n} \frac{e^{ika(m-n)}}{|n-m|^3} \ .
\end{aligned}
\tag{II.2.82}
$$

The sum can be converted to

$$
\sum_{m=1}^{N/2} \frac{2}{m^3} \cos kam \ .
\tag{II.2.83}
$$

Therefore we obtain the local field at point $n$ to be

$$
E_n = \left( \frac{4p^*}{a^3} \sum_{m=1}^{N/2} \frac{1}{m^3} \cos kam \right) e^{ikan} e^{-i\omega t} \ .
\tag{II.2.84}
$$

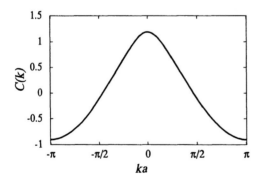

**Fig. II.2.9.** The function $C(k)$, defined by Eq. II.2.86, in the $N \to \infty$ limit.

From the equation of motion we obtain

$$
-\omega^2 = -\omega_0^2 + \frac{4\alpha\omega_0^2}{a^3} C(k) \ ,
\tag{II.2.85}
$$

$$
\text{where} \quad C(k) = \sum_{m=1}^{N/2} \frac{1}{m^3} \cos kam \ .
\tag{II.2.86}
$$

The function $C(k)$, plotted in Figure II.2.9, looks like a distorted cosine curve. Assuming $C(k)$ is known, we can solve for $\omega$:

$$
\omega = \omega_0 \sqrt{1 - \frac{4\alpha}{a^3} C(k)} \ .
\tag{II.2.87}
$$

The result is plotted in Figure II.2.10. For small polarizability the square root can be expanded and the dispersion is small: $\omega = \omega_0(1 - (2\alpha/a^3)C(k))$. As the coefficient $\gamma = 4\alpha/a^3$ increases the $k = 0$ mode frequency approaches zero. If the polarizability exceeds a critical value, $\alpha_{\text{crit}}$, the solution for small $k$ becomes imaginary. In this range of parameters the system becomes *ferroelectric*: A spontaneous polarization develops along the chain, and the inversion symmetry is broken.[3] In a more general context, symmetry breaking soft modes are called *Goldstone bosons*.

Our solution relies on the assumption that the polarization is oscillating around $p = 0$ (see Eq. II.2.80). For $\alpha > \alpha_{\text{crit}}$ the anharmonic terms in Eq. I.2.17 has to be taken into account, and the vibrational frequencies can be derived in terms of expanding $p$ around the new equilibrium polarization.

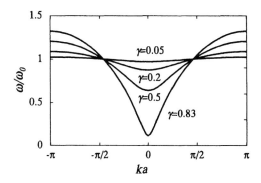

**Fig. II.2.10.** The frequency as a function of wavenumber for several values of $\gamma = 4\alpha/a^3$.

The softening of an optical phonon mode is a typical precursor of the ferroelectric transition, and the soft optical phonons has been observed in ferroelectric materials like $SrTiO_3$.

## 2.16 Hint: Soft Phonons Again

The solution is similar to that of Problem 2.15, except the softening of the mode will occur at the Brillouin zone boundary, $k = \pm\pi/a$. For large values of the coupling constant $\alpha$ the direction of dipole moments alternates along the chain; the static order is anti-ferroelectric.

---

[3] This is discussed in Problems 9.1 and 9.2.

# 3 Electronic Band Structure

## 3.1 Solution: Nearly Free Electrons in One Dimension

(a) In the NFE approximation the periodic potential is treated as a perturbation relative to the kinetic energy of the electrons. Consequently, the potential energy must be much smaller than the kinetic energy: $|V_0| \ll (\hbar\pi/a)^2/2m$, $|V_1| \ll (\hbar 2\pi/a)^2/2m$, and so on. The energy bands will be essentially parabolic, with gaps at every point of degeneracy.

(b) The degeneracy of energies is removed by mixing the corresponding wavefunctions and the perturbation potential. The matrix element is $E_{12} = \langle\psi_1|V(x)|\psi_2\rangle$. In the neighborhood of $k = \pi/a$ the normalized wavefunctions are $\psi_1 = \frac{1}{\sqrt{L}}e^{ikx}$ (for the lower band) and $\psi_2 = \frac{1}{\sqrt{L}}e^{i(k-2\pi/a)x}$ for the upper band ($L$ is the length of the system). (The wavefunction for the upper band can be deduced from the "extended zone" scheme.)

$$
\begin{aligned}
E_{12} &= \frac{1}{L}\int_0^L e^{-ikx}V(x)e^{i(k-2\pi/a)x}\,dx \\
&= \frac{1}{L}\int_0^L V(x)e^{-2\pi i/ax}\,dx = V_1/2 .
\end{aligned}
\tag{II.3.1}
$$

For the band splitting at $k = 0$ the calculation is similar, except the wavefunctions are $\psi_1 = \frac{1}{\sqrt{L}}e^{i(k-2\pi/a)x}$ and $\psi_2 = \frac{1}{\sqrt{L}}e^{i(k+2\pi/a)x}$. This leads to the Fourier transform

$$
E_{12}(k=0) = \frac{1}{L}\int_0^L V(x)e^{-4\pi i/ax}dx = V_2/2 .
\tag{II.3.2}
$$

In both cases, diagonalizing the Hamiltonian matrix yields the gap, $\Delta E = 2E_{12}$. The bands are illustrated in Figure II.3.1. Note that there is no gap at $k = 0$ for a simple "harmonic" potential ($V_2 = 0$).

## 3.2 Solution: Nearly Free Electrons in Dirac-Delta Potentials

Similar to the solution of Problem 3.1, the energy gaps are determined by the Fourier components of the lattice potential. For Dirac-delta potentials

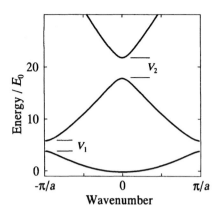

**Fig. II.3.1.** Electronic bands in the nearly free-electron approximation. The energy scale is in units of $E = \hbar^2/2m$.

the matrix elements are

$$V_{G=2\pi n/a} = E_{12} = \frac{1}{L} \int_0^L V(x)e^{-i2\pi nx/a} = V_0 \qquad \text{(II.3.3)}$$

independent of $n$. Therefore all gaps are the same, $\Delta = 2V_0$.

## 3.3 Solution: Tight-Binding in Dirac-Delta Potentials

(a) The wavefunction of a single atom will have the form

$$\psi = Ae^{-\kappa|x|} , \qquad \text{(II.3.4)}$$

with its energy being given by $E = -\frac{\hbar^2}{2m}\kappa^2$. Around $x = 0$, we have

$$-\frac{\hbar^2}{2m}\left(\left.\frac{\partial\psi}{\partial x}\right|_+ - \left.\frac{\partial\psi}{\partial x}\right|_-\right) + \psi aV_0 = 0 \qquad \text{(II.3.5)}$$

$$\left.\frac{\partial\psi}{\partial x}\right|_+ = -\kappa Ae^{-\kappa x} = -\kappa A \qquad \text{(II.3.6)}$$

$$\left.\frac{\partial\psi}{\partial x}\right|_- = +\kappa A . \qquad \text{(II.3.7)}$$

Therefore,

$$\frac{\hbar^2}{2m}2\kappa A + AaV_0 = 0 \quad \longrightarrow \quad \kappa = -\frac{aV_0 m}{\hbar^2} . \qquad \text{(II.3.8)}$$

$V_0$ must therefore be negative. Finally, from $\int |\psi|^2 \, dx = 1$, we can see that $A = \sqrt{\kappa}$.

(b) In the tight-binding approximation, $a\kappa \gg 1$. The energy as a function of wavenumber is given by Eq. I.3.9:

$$E(k) = E_0 - \frac{\beta + 2\gamma\cos ka}{1 + 2\alpha\cos ka}, \qquad \text{(II.3.9)}$$

where

$$\alpha = \int dx \psi^\dagger(x)\psi(x-a) \tag{II.3.10}$$

$$\gamma = -aV_0 \int dx \psi^\dagger(x)\delta(x-a)\psi(x-a) -$$

$$-aV_0 \int dx \psi^\dagger(x)\delta(x+a)\psi(x-a) - \tag{II.3.11}$$

$$-aV_0 \int dx \psi^\dagger(x)\delta(x \pm 2a)\psi(x-a) - \dots$$

For Dirac-delta potentials the integrand in the expression for $\beta$ is exactly zero. Using $\psi = \sqrt{\kappa}e^{-\kappa|x|}$, we obtain

$$\alpha = \kappa \left[ \int_{-\infty}^{0} e^{2\kappa x}e^{-\kappa a}\,dx + \int_{0}^{a} e^{-\kappa a}dx + \int_{a}^{\infty} e^{-2\kappa x}e^{-\kappa a}\,dx \right]$$

$$= \kappa a e^{-\kappa a} + e^{-\kappa a}\left(\frac{1}{2} + e^{-2\kappa a}\right) \tag{II.3.12}$$

$$\gamma = -\kappa a V_0 e^{-\kappa a} + \dots . \tag{II.3.13}$$

We only kept the dominant term for $a\kappa \gg 1$. Note that $\alpha \ll 1$, and therefore, from Eq. II.3.9,

$$E(k) = E_0 - 2\gamma \cos ka \tag{II.3.14}$$

$$= E_0 + 2V_0 a\kappa e^{-\kappa a} \cos ka . \tag{II.3.15}$$

With $V_0 < 0$ this energy band has a minimum at $k = 0$. The bandwidth is

$$W = 4V_0 a\kappa e^{-\kappa a} . \tag{II.3.16}$$

And finally since $\kappa = -\frac{aV_0 m}{\hbar^2}$, this becomes

$$W = 4\frac{(aV_0)^2 m}{\hbar^2}e^{-\kappa a} , \tag{II.3.17}$$

in agreement with the general solution, Eq. II.3.42.

## 3.4 Solution: Dirac-Delta Potentials

(a) We place atoms at the $x = na$ positions and first look at the $0 < x < a$ interval. Between the atoms, the Schrödinger equation is

$$\frac{-\hbar}{2m}\frac{\partial^2 \psi}{\partial x^2} = E\psi , \tag{II.3.18}$$

$$\psi = Ae^{iKx} + Be^{-iKx} , \tag{II.3.19}$$

where $K = \sqrt{\frac{2mE}{\hbar^2}}$. As we will see, depending on the sign of $V_0$, $K$ can be real or imaginary.

To extend the solution to the full $-\infty < x < \infty$ range we use Bloch's theorem,

$$\psi = e^{-ikx} u_k(x) , \qquad (\text{II.3.20})$$

where $u_k(x)$ is a periodic function. Combining this with Eq. II.3.19 we obtain, in the $0 < x < a$ range,

$$u_k(x) = A e^{i(K-k)x} + B e^{-i(K+k)x} \qquad \text{for } 0 < x < a . \qquad (\text{II.3.21})$$

The full $u_k(x)$ function can be generated by a periodic repetition of this function. At $x = na$ the wavefunction is continuous but its derivative is not. The jumps in the derivative can be found by integrating the Schrödinger equation over a small range around $a$ $(a - \delta < x < a + \delta)$. This yields,

$$\psi|_{x=a+\delta} = \psi|_{x=a-\delta}$$

$$\text{and} \quad \left.\frac{\partial \psi}{\partial x}\right|_{x=a+\delta} - \left.\frac{\partial \psi}{\partial x}\right|_{x=a-\delta} = \frac{2maV_0}{\hbar^2}\psi(a) . \qquad (\text{II.3.22})$$

Now we can use the periodic nature of $u_k(x)$: First since $e^{ikx}$ is a continuous function, Eq. II.3.22 leads to $u_k(a + \delta) = u_k(a - \delta)$. But in the $\delta \to 0$ limit, $u_k(a + \delta) = u_k(a)$, and from Eq. II.3.21 we obtain

$$A + B = A e^{i(K-k)a} + B e^{-i(K+k)a} . \qquad (\text{II.3.23})$$

Second, the derivative of $\psi$ is

$$\frac{\partial \psi}{\partial x} = ik e^{ikx} u_k(x) + e^{ikx}\frac{du_k(x)}{dx} , \qquad (\text{II.3.24})$$

and from Eq. II.3.22 we obtain

$$e^{ik(a+\delta)}\left.\frac{du_k(x)}{dx}\right|_{a+\delta} - e^{ik(a-\delta)}\left.\frac{du_k(x)}{dx}\right|_{a-\delta} = \frac{2maV_0}{\hbar^2}e^{ika}u_k(x)$$

$$\left.\frac{du_k(x)}{dx}\right|_0 - \left.\frac{du_k(x)}{dx}\right|_a = \frac{2maV_0}{\hbar^2}u_k(0) . \qquad (\text{II.3.25})$$

From Eq. II.3.21 we can explicitly calculate the derivative of $u_k(x)$:

$$\frac{du_k(x)}{dx} = i(K - k)A e^{i(K-k)x} - i(K + k)B e^{-i(K+k)x} . \qquad (\text{II.3.26})$$

Putting this together with Eq. II.3.25, we obtain the equation

$$i(K - k)A - i(K + k)B - i(K - k)A e^{i(K-k)a}$$

$$+ i(K + k)B e^{-i(K+k)a} = \frac{2maV_0}{\hbar^2}(A + B) . \qquad (\text{II.3.27})$$

This equation, together with Eq. II.3.23, gives us two linear equations for the two unknowns, $A$ and $B$. We begin by rewriting Eq. II.3.23 and then inserting Eq. II.3.27:

$$A\left(1 - e^{i(K-k)a}\right) + B\left(1 - e^{-i(K+k)a}\right) = 0 \tag{II.3.28}$$

$$A\left[i(K-k)\left(1 - e^{i(K-k)a}\right) - \frac{2maV_0}{\hbar^2}\right]$$

$$+B\left[i(K+k)\left(e^{i(K+k)a} - 1\right) - \frac{2maV_0}{\hbar^2}\right] = 0 . \tag{II.3.29}$$

We introduce the new variable,

$$Z = A\left(1 - e^{i(K-k)a}\right) = B\left(e^{i(K+k)a} - 1\right) , \tag{II.3.30}$$

and obtain

$$i2KZ - \frac{2maV_0}{\hbar^2}(A + B) = 0 . \tag{II.3.31}$$

However we also know that

$$A + B = \frac{Z}{1 - e^{i(K-k)a}} + \frac{Z}{e^{i(K+k)a} - 1} . \tag{II.3.32}$$

To find a nonzero $Z$, we need to satisfy

$$i2K\left(1 - e^{i(K-k)a}\right)\left(e^{i(K+k)a} - 1\right)$$

$$-\frac{2maV_0}{\hbar^2}\left[1 - e^{i(K-k)a} + e^{i(K+k)a} - 1\right] = 0 . \tag{II.3.33}$$

Multiplying this equation by $e^{ika}$ results in[1]

$$i2K\left[2\cos Ka - 2\cos ka\right] + \frac{2maV_0}{\hbar^2}2i\sin Ka = 0 \tag{II.3.34}$$

$$\cos ka = \frac{1}{2K}\frac{2maV_0}{\hbar^2}\sin Ka + \cos Ka . \tag{II.3.35}$$

Equation I.3.21 comes from this. The parameter $\alpha$ is therefore

$$\alpha = \frac{ma}{\hbar^2} . \tag{II.3.36}$$

To summarize part (a), we obtained the relation $\cos ka = \frac{\kappa}{K}\sin Ka + \cos Ka$ with $\kappa = \frac{maV_0}{\hbar^2}$ and the energy $E = \frac{\hbar^2 K^2}{2m}$. This is valid for $V_0 > 0$. When $V_0 < 0$, we get a negative energy. Introducing $\kappa^* = -\frac{maV_0}{\hbar} > 0$ and $K = i\Gamma$, the energy is then $E = -\frac{\hbar^2\Gamma^2}{2m}$ and

$$\cos ka = -\frac{\kappa^*}{\Gamma}\sinh \Gamma a + \cosh \Gamma a . \tag{II.3.37}$$

---

[1] In Kittel's words [2] p. 192 : "It is rather tedious to obtain this equation."

(b) Let us now look at the nearly free-electron limit at $k = \frac{\pi}{a}$ and around the energy of $E_0 = \frac{\hbar^2}{2m}\left(\frac{\pi}{a}\right)^2$. Assume that $K = K_0 + \nu$, where $K_0^2 = \frac{2mE_0}{\hbar^2} = \left(\frac{\pi}{a}\right)^2 \rightarrow K_0 = \pm\frac{\pi}{a}$. We will use Eq. II.3.35 to determine $\nu$ for $K_0 = \pi/a$. Inserting $ka = \pi$, $\sin Ka \approx -\nu a$ and $\cos Ka \approx -1 + \frac{1}{2}(\nu a)^2$, and replacing $K$ by $\pi/a$ in the denominator of the first term yields

$$-1 = \frac{\kappa}{\pi/a}(-\nu a) - 1 + \frac{1}{2}(\nu a)^2 . \qquad (II.3.38)$$

The solutions are $\nu = 2\kappa/\pi a$ and $\nu = 0$. Therefore, to first-order in $\nu$, the energies of the two branches are $E_1 = E_0$ and $E_2 = E_0 + \frac{\hbar^2}{m}\frac{\pi}{a}\nu = E_0 + 2V_0$.

The nearly free-electron approximation (see Problem 3.2) yields $\Delta = 2V_0$, in agreement with the result here. However, perturbation theory suggests that in the neighborhood of $k = \pm\pi/a$ the gap develops by a downshift of the lower-energy band, and an upshift of the upper-energy band. The exact solution tells us that switching the potential raises the lower band energy at every wavenumber so that the energy at $k = \pm\pi/a$ does not change.[2]

(c) In the tight-binding limit, we have to take a negative $V_0$ (otherwise there would be no bound state in the single atom potential). We will want to calculate the bandwidth, defined as $W = E(\frac{\pi}{a}) - E(0) = \frac{\hbar^2}{2m}(\Gamma_2^2 - \Gamma_1^2)$, where $\Gamma_1$ and $\Gamma_2$ are the solutions of Eq. II.3.37 for $k = 0$ and $k = \pi/a$, respectively.

If the potential is strong, then $\kappa^* \gg 1/a$. The left hand side of Eq. II.3.37 is bounded between $\pm 1$. The only way to make the right hand side small enough is to have $\Gamma \gg 1/a$. In this limit the $\sinh x$ and $\cosh x$ functions can be replaced by $\frac{1}{2}\exp x$. For $k = 0$ Eq. II.3.37 gives

$$1 = -\frac{\kappa^*}{\Gamma_1}\sinh \Gamma_1 a + \cosh \Gamma_1 a \approx \frac{1}{2}e^{\Gamma_1 a}\left(1 - \frac{\kappa^*}{\Gamma_1}\right) . \qquad (II.3.39)$$

For $k = \frac{\pi}{a}$,

$$-1 = -\frac{\kappa^*}{\Gamma_2}\sinh \Gamma_2 a + \cosh \Gamma_2 a \approx \frac{1}{2}e^{\Gamma_2 a}\left(1 - \frac{\kappa^*}{\Gamma_2}\right) . \qquad (II.3.40)$$

By rearranging the terms we get $\Gamma_1 = \kappa^* - 2\Gamma_1 e^{-\Gamma_1 a}$ and $\Gamma_2 = \kappa^* + 2\Gamma_2 e^{-\Gamma_2 a}$. Due to the exponential factor $\Gamma_1 \approx \Gamma_2 = \kappa^*$. The small difference between the two quantities determines the bandwidth:

$$W = \frac{\hbar^2}{2m}(\Gamma_2^2 - \Gamma_1^2) = \frac{\hbar^2}{2m}2\Gamma(\Gamma_2 - \Gamma_1) , \qquad (II.3.41)$$

where $\Gamma \approx \kappa^*$ is the average of $\Gamma_1$ and $\Gamma_2$, and $\Gamma_2 - \Gamma_1 = 4\Gamma e^{-\Gamma a}$. The bandwidth is

$$W = \frac{\hbar^2}{2m}8\Gamma^2 e^{-\Gamma a} = 4\frac{(aV_0)^2 m}{\hbar^2}e^{-\kappa^* a} . \qquad (II.3.42)$$

This result agrees with the bandwidth obtained in the tight-binding approximation, Solution 3.3.

---

[2] for illustration, see Kittel [2] p. 193.

## 3.5 Solution: Band Overlap

The Schrödinger equation in one dimension is an ordinary second-order differential equation. (In contrast, for two or three dimensions we have partial differential equations.) Second-order differential equations may only have two independent solutions for a given set of parameters (i.e., for fixed energy). According to the Bloch theorem, the wavenumber $k$ is a good quantum number. As long as time reversal symmetry or inversion symmetry holds the energy at $k$ must be equal to the energy at $-k$. There are no more independent solutions allowed for the same energy.

## 3.6 Solution: Nearly Free Electrons in Two Dimensions

We have to solve Eq. I.3.6, with $k = \frac{\pi}{a}(\frac{1}{2}, \frac{1}{2})$. The reciprocal lattice is drawn in Figure II.3.2 (in units of $\frac{2\pi}{a}$). In the vicinity of the $L = (\frac{1}{2}, \frac{1}{2})$ point, the energy $E(k - G)$ is degenerate for $G = (0,0)$, $(0,1)$, $(1,0)$, and $(1,1)$. Therefore the possible $G - G'$ values in Eq. I.3.6 are $(0,\pm1)$, $(\pm1,0)$, $(\pm1,\pm1)$, and $(0,0)$ and the corresponding values of $V_{G-G'}$ are

$$V_{(0,\pm1)} = V_{(\pm1,0)} \equiv V_{10} \qquad \text{(II.3.43)}$$
$$V_{(1,1)} = V_{(-1,-1)} \equiv V_{11} \qquad \text{(II.3.44)}$$
$$V_{(0,0)} \equiv V_{00}. \qquad \text{(II.3.45)}$$

$k_y$

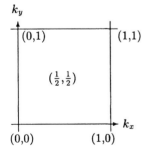

Fig. II.3.2. Reciprocal space of a square lattice.

We will now introduce the notations

$$\frac{\hbar^2}{2m}k - G_{i,j}^2 = E_{ij} \quad \text{and} \quad C_{k-G_{i,j}} = c_{ij}. \qquad \text{(II.3.46)}$$

The set of equations to solve are

$$(E - E_{00})c_{00} - V_{00}c_{00} - V_{10}c_{01} - V_{10}c_{10} - V_{11}c_{11} = 0 \quad \text{(II.3.47)}$$
$$(E - E_{10})c_{10} - V_{10}c_{00} - V_{11}c_{01} - V_{00}c_{10} - V_{10}c_{11} = 0 \quad \text{(II.3.48)}$$
$$(E - E_{01})c_{01} - V_{10}c_{00} - V_{00}c_{01} - V_{11}c_{10} - V_{10}c_{11} = 0 \quad \text{(II.3.49)}$$
$$(E - E_{11})c_{11} - V_{11}c_{00} - V_{10}c_{01} - V_{10}c_{10} - V_{00}c_{11} = 0 , \quad \text{(II.3.50)}$$

or we can rewrite this in matrix form (where $V_E = V_{00} - E$):

$$\begin{pmatrix} E_{00} + V_E & V_{10} & V_{10} & V_{11} \\ V_{10} & E_{10} + V_E & V_{11} & V_{10} \\ V_{10} & V_{11} & E_{01} + V_E & V_{10} \\ V_{11} & V_{10} & V_{10} & E_{11} + V_E \end{pmatrix} \begin{pmatrix} c_{00} \\ c_{10} \\ c_{01} \\ c_{11} \end{pmatrix} = 0 .$$

(II.3.51)

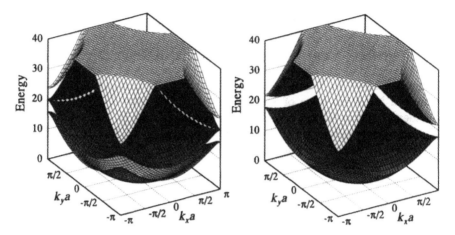

**Fig. II.3.3.** Nearly free-electron bands for $V_{10} = 2$ and $V_{11} = 0$ (left panel), and for $V_{10} = 0$ and $V_{11} = 2$ (right panel).

At the $(\frac{1}{2}, \frac{1}{2})$ point we have $E_{00} = E_{01} = E_{10} = E_{11}$. In general, we can solve this equation to get four solutions for $E$ with the condition det( ) = 0. However, in this case we want to consider two special cases. First, when $V_{11} = 0$. Introducing $\delta E = E_{00} + V_{00} - E$, Eq. II.3.51 becomes

$$\begin{vmatrix} \delta E & V_{10} & V_{10} & 0 \\ V_{10} & \delta E & 0 & V_{10} \\ V_{10} & 0 & \delta E & V_{10} \\ 0 & V_{10} & V_{10} & \delta E \end{vmatrix} = 0 .$$

(II.3.52)

This determinant is solved, obtaining $\delta E^2 (\delta E^2 - 4V_{10}^2) = 0$; therefore

$$\delta E = 0 \quad \text{and} \quad \delta E = \pm\sqrt{2}V_{10}$$

(II.3.53)

are solutions. The degeneracy is not removed completely ($\delta E = 0$), so the gap is zero.

The second case is for $V_{01} = 0$, where Eq. II.3.51 becomes

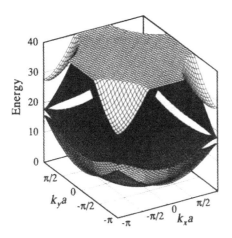

**Fig. II.3.4.** Nearly free-electron bands for $V_{10} = 3$ and $V_{11} = 2$.

$$\begin{vmatrix} \delta E & 0 & 0 & V_{11} \\ 0 & \delta E & V_{11} & 0 \\ 0 & V_{11} & \delta E & 0 \\ V_{11} & 0 & 0 & \delta E \end{vmatrix} = 0 \ . \tag{II.3.54}$$

This leads to $(\delta E^2 - V_{11}^2)^2 = 0$, which has solutions of

$$\delta E = \pm V_{11} \ . \tag{II.3.55}$$

Here the gap is $\Delta = 2V_{11}$.

In Figures II.3.3–II.3.4 the bands for various values of $V_{10}$ and $V_{11}$ are plotted.

## 3.7 Solution: Nearly Free Electron Bands

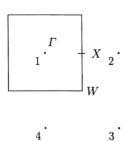

**Fig. II.3.5.** Unit cell in reciprocal space with the $\Gamma$, $W$, and $X$ points labeled. The numbered points indicate the centers of the next-neighbor Brillouin zones.

The reciprocal space cell for this problem is drawn in Figure II.3.5. For a weak potential the energy dispersion of the electrons remains close to quadratic. Let us place paraboloids at the points labeled 1, 2, 3, and 4 in Figure II.3.5.

We can then draw these bands in a plot of $E(k)$ along the $\Gamma - W - X$ lines, shown in Figure II.3.6 with the curves labeled for which paraboloid each came from.

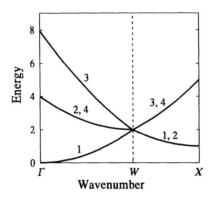

**Fig. II.3.6.** $E(k)$ as a function of $k$ along the $\Gamma - W$ and $W - X$ lines

When the lattice potential is turned on, the degeneracies split, but the energies will remain close to the values shown in Figure II.3.6.

## 3.8 Solution: Instability at the Fermi Wavenumber

According to Eq. I.3.7,

$$E^{\pm}(k) = \frac{1}{2}(E_k + E_{k-G}) \pm \frac{1}{2}\sqrt{(E_k - E_{k-G})^2 + 4V_G^2} , \qquad (\text{II.3.56})$$

where $E_k = \frac{\hbar^2}{2m}k^2$. From the periodic potential given for this problem, $V(x) = V_0 \cos qx$, we can calculate

$$V_G = \frac{1}{L}\int_0^L V(x)e^{iGx} = \frac{1}{2}V_0 , \qquad (\text{II.3.57})$$

where $G = 2k_F$ and the gap opens at the $k = \pm k_F$ points, as sketched in Figure II.3.7.

Since the band is symmetric, we can simply work with the $k > 0$ portion and then multiply the result by 2. We write $E^- = E^{(0)} - \Delta E$, where $E^{(0)} = \frac{\hbar^2}{2m}k^2$ and

$$\Delta E = \frac{\hbar^2}{2m}\frac{1}{2}\left\{(k-q)^2 - k^2 - \sqrt{[k^2 - (k-q)^2]^2 + 4V_0^2 \left(\frac{2m}{\hbar^2}\right)^2}\right\} \qquad (\text{II.3.58})$$

Defining the new variable, $k' = \frac{1}{2}q - k$, we obtain

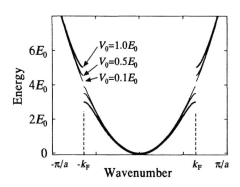

**Fig. II.3.7.** Band-gap opening for a periodic potential applied to a one-dimensional system. $E_0 = \hbar^2/ma^2$.

$$\Delta E = \frac{\hbar^2}{2m}\left\{ qk' - \sqrt{(qk')^2 + V_0^2\left(\frac{2m}{\hbar^2}\right)^2}\right\}. \tag{II.3.59}$$

For a full band at zero temperature, the total energy is

$$E_{\text{tot}} = 2\int_0^{q/2} \frac{dk}{2\pi/L}(E^0 - \Delta E) = E_{\text{tot}}^0 - \Delta E_{\text{tot}}, \tag{II.3.60}$$

where the 2 is to account for both $k > 0$ and $k < 0$, and

$$E_{\text{tot}}^{(0)} = \frac{L}{\pi}\int_0^{q/2} \frac{\hbar^2}{2m}k^2 dk = \frac{L\hbar^2}{24\pi m}q^3 \tag{II.3.61}$$

is the energy of the unperturbed band. $\Delta E_{\text{tot}}$ is given by

$$\Delta E_{\text{tot}} = \frac{L}{\pi}\int_0^{q/2} \frac{\hbar^2}{2m}\left\{ qk' - \sqrt{(qk')^2 + V_0^2\left(\frac{2m}{\hbar^2}\right)^2}\right\} dk'. \tag{II.3.62}$$

This integration results in (using the fact that the periodic potential is weak, i.e., $V_0 \ll \frac{\hbar^2}{2m}(\frac{q}{2})^2$)

$$\Delta E_{\text{tot}} = \frac{L}{\pi}\frac{V_0^2 m}{\hbar^2 q^2}\log\left(\frac{V_0}{\frac{\hbar^2 q^2}{2m}}\right). \tag{II.3.63}$$

Note that the dependence on the perturbation, $V_0$, is $\Delta E_{\text{tot}} \sim V_0^2 \log V_0$; therefore for small $V_0$ the correction will be negative, $\Delta E_{\text{tot}} < 0$. This means that the total energy of the electrons is reduced when a weak periodic potential is applied! The reason for the decrease is that the energy of the *occupied* states becomes lower, while the energy of the *empty* states rises. Ultimately this effect leads to an instability in the one-dimensional electron gas (the Peierls instability), as discussed in Problem 9.6.

## 3.9 Solution: Electrons in 2D Nearly Free Electron Band

(a) With one electron per site, the band becomes half-filled and will therefore
lead to a metal. The Fermi surface is sketched by taking a constant energy
cut such that the area within the Fermi line is half of the area of the Brillouin
zone (see left panel of Figure II.3.8). Since the Fermi line does not get too
close to the Brillouin zone boundary, in good approximation the cut is a
circle. Note the difference between this Fermi surface and the Fermi surface
in the tight-binding approximation, Figures II.3.11 and II.3.13. The Fermi
wavevector is

$$\pi k_F^2 = \frac{1}{2} \left( \frac{2\pi}{a} \right)^2 \quad \longrightarrow \quad k_F = \frac{\sqrt{2\pi}}{a} \ . \tag{II.3.64}$$

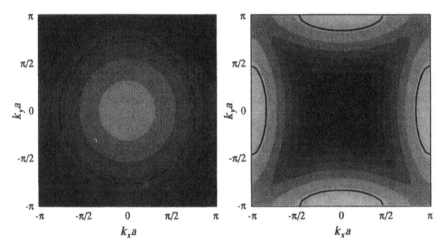

**Fig. II.3.8.** Left panel: Contour plot of the energy surface for the lower-energy
band. Darker shades correspond to higher energies. The dashed line is the Fermi
surface for one electron per site; the solid line is that for two electrons per site.
Right panel: Contour plot for the upper band. The solid line is the Fermi surface
for two electrons per site.

(b) With two electrons per site, assuming that the constant energy lines
are circles, we obtain

$$\pi k_F^2 = \left( \frac{2\pi}{a} \right)^2 \quad \longrightarrow \quad k_F = 2\frac{\sqrt{\pi}}{a} \ . \tag{II.3.65}$$

This value is larger than $\frac{\pi}{a}$, therefore we have to consider the two bands, as
illustrated in Figure II.3.9. There will be two Fermi surfaces, one in the upper
band and another one in the lower band. The Fermi line in the upper band
is shown in the right panel of Figure II.3.8. Each of the bands is partially

filled; with two electrons per site the NFE approximation leads to a metal. In contrast, the tight-binding model (Problem 3.12) results in an insulator.

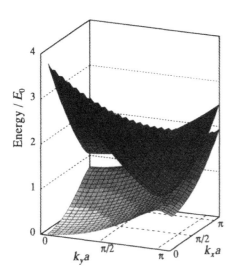

## 3.10 Solution: Square Lattice

The Schrödinger equation is

$$-\frac{\hbar^2}{2m}\frac{\delta^2\psi}{\delta k^2} + V(r)\psi = E\psi. \tag{II.3.66}$$

If the potential is weak, the solutions will be plane waves:

$$\psi(k^*) = \frac{1}{\sqrt{V}}e^{ik^*r} \quad \text{and} \quad E_k = \frac{\hbar^2|k^*|^2}{2m}, \tag{II.3.67}$$

where $k^*$ extends over the entire $k$-space (see Figure II.3.10).

We can transform the $k^*$ wavevector into the first Brillouin zone by

$$k^* = k + nb_1 + mb_2, \tag{II.3.68}$$

where $k$ is the reciprocal lattice vector within the first Brillouin zone, $b_1$ and $b_2$ are the primitive reciprocal lattice vectors, and $n$ and $m$ are integers. The primitive reciprocal lattice vectors for this lattice are given by

$$b_1 = \begin{pmatrix} 1.256 \text{ Å}^{-1} \\ 0 \end{pmatrix}, \quad b_2 = \begin{pmatrix} 0 \\ 1.256 \text{ Å}^{-1} \end{pmatrix}. \tag{II.3.69}$$

Of course there will be many energies for a given $k$ and there may be more than one $u(r)$ solution for each energy. The length of the $k^*$ vector is shown

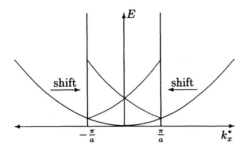

Fig. II.3.10. Zone folding into the first Brillouin zone.

for several values of $n$ and $m$ (with the value of $k = 0.5$ Å$^{-1}$ given in the problem) in Table II.3.1. Since the energy increases with $|k^*|$, the three lowest energies are (a) $E = 0.95$ eV ($n = 0$, $m = 0$); (b) $E = 2.17$ eV ($n = -1$, $m = 0$) and (c) $E = 6.96$ eV ($n = 0$, $m = \pm 1$).
    From

$$e^{ik^* \cdot r} = e^{ikr} u(r), \quad r = \begin{pmatrix} x \\ y \end{pmatrix}, \tag{II.3.70}$$

we can solve for $u(r)$:

$$\text{(a)} \qquad u(r) = 1, \tag{II.3.71}$$

$$\text{(b)} \qquad u(r) = e^{-ib_1 x} = e^{-i1.256x}, \tag{II.3.72}$$

$$\text{(c)} \qquad u(r) = e^{\pm ib_2 y} = e^{\pm i1.256y}. \tag{II.3.73}$$

Note that the functions $u(r)$ have the periodicity of the lattice, $u(r) = u(r + R)$. The third energy level is degenerate; there are two corresponding wavefunctions.

Table II.3.1. Magnitudes of $k^*$ for Several Values of $n$ and $m$

| $n$ | $m$ | $|k^*|$ |
|---|---|---|
| 0 | 0 | 0.5 |
| −1 | 0 | 0.756 |
| 1 | 0 | 1.756 |
| 0 | 1 | 1.351 |
| 0 | −1 | 1.351 |
| 1 | 1 | 2.159 |

## 3.11 Solution: Tight-Binding Band in Two Dimensions

(a) In Figure II.3.11 we have plotted the lowest band of this two-dimensional square lattice

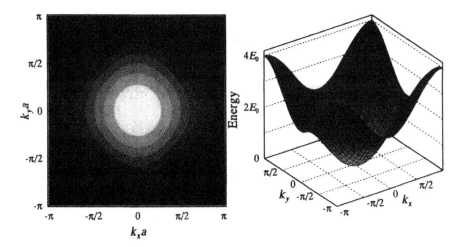

**Fig. II.3.11** Lowest tight-binding band of a two-dimensional square lattice.

(b) To show that the constant energy lines are perpendicular to the Brillouin zone boundary, we must show that

$$\frac{dk_{xy}}{dk_x} = 0 \quad \text{for } k_x = \frac{\pi}{a} \text{ and } E = \text{constant.} \tag{II.3.74}$$

We have

$$dE = 0 = \frac{\partial E}{\partial k_x}\, dk_x + \frac{\partial E}{\partial k_y}\, dk_y \;, \tag{II.3.75}$$

and hence

$$\frac{dk_x}{dk_y} = -\frac{\frac{\partial E}{\partial k_y}}{\frac{\partial E}{\partial k_x}} = -\frac{\sin k_x}{\sin k_y} \;. \tag{II.3.76}$$

For $k_x = \pi$ this equation is zero except for the cases when $k_y = 0$ or $k_y = \pi$. An analogous result is obtained for $dk_y/dk_x$ at $k_y = \pi/a$.

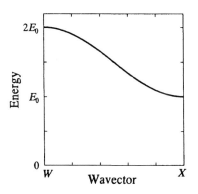

**Fig. II.3.12.** $E(k)$ for a two-dimensional square lattice along the $W - X$ line.

(c) Figure II.3.11 helps to sketch the energy as a function of $k$ along the $W - X$ line, shown in Figure II.3.12.

## 3.12 Solution: Electrons in 2D Tight-Binding Band

We will use the results of Problem 3.11.

(a) One electron per site means a half-filled band (due to the two spin states). The Fermi surface is the constant energy cut where the area in $k$-space is *half* of the total area of the Brillouin zone. This is sketched in Figure II.3.13. This is a metal since the band is partially filled.

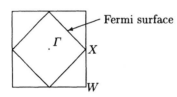

**Fig. II.3.13.** The Fermi surface for a square lattice with a half-filled band.

(b) Two electrons per site leads to a complete filling of the lowest band. In this case the Fermi surface *disappears* and the material becomes an insulator. To demonstrate this, we sketch a nearly full band in Figure II.3.14. The area of the segments at the corners goes to zero as the band filling goes to two electrons per site (full band).

**Fig. II.3.14.** The Fermi surface for a square lattice with a nearly filled band.

## 3.13 Solution: Dirac-Delta Potentials in Two Dimensions

(a) In the nearly free-electron approximation we need the potential to be weak such that

$$V_G \ll \frac{\hbar^2 k^2}{2m} , \tag{II.3.77}$$

where $V_G = V_0$ is the Fourier component of the lattice potential at the reciprocal lattice vector $G$. The Fourier components of a Dirac-delta potential are all $V_0$. This leads to a condition for $V_0$,

$$V_0 \ll \frac{\hbar^2}{2m} \frac{\pi^2}{a^2} \ , \qquad (\text{II.3.78})$$

where we have used $k \approx \frac{\pi}{a}$.

(b) The lower and upper bands will be separated by a finite energy gap at any fixed value of wavenumber $k$. However, if the lattice potential is weak, then in two or more dimensions the bands will overlap and there is no energy gap. For illustration, see Solutions 3.6 or 3.7.

## 3.14 Hint: Effective Mass

The effective mass is defined in Eqs. I.3.16 and I.3.17. The electron energy in the tight-binding approximation is

$$E(\boldsymbol{k}) = -E_1 (\cos k_x a + \cos k_y a + \cos k_z a) \ , \qquad (\text{II.3.79})$$

with a bandwidth of $6E_1$. The derivatives of $E(\boldsymbol{k})$ are

$$\frac{\partial E}{\partial k_x} = E_1 a \sin k_x a \qquad (\text{II.3.80})$$

$$\frac{\partial E}{\partial k_y} = E_1 a \sin k_y a \qquad (\text{II.3.81})$$

$$\frac{\partial E}{\partial k_z} = E_1 a \sin k_z a \qquad (\text{II.3.82})$$

$$\frac{\partial^2 E}{\partial k_x^2} = E_1 a^2 \cos k_x a \qquad (\text{II.3.83})$$

$$\frac{\partial^2 E}{\partial k_y^2} = E_1 a^2 \cos k_y a \qquad (\text{II.3.84})$$

$$\frac{\partial^2 E}{\partial k_x \partial k_y} = 0 \ \ldots \qquad (\text{II.3.85})$$

Accordingly, the effective mass tensor is

$$\mathsf{M} = \hbar^2 \begin{pmatrix} E_1 a^2 \cos k_x a & 0 & 0 \\ 0 & E_1 a^2 \cos k_y a & 0 \\ 0 & 0 & E_1 a^2 \cos k_z a \end{pmatrix}^{-1} . \qquad (\text{II.3.86})$$

The three reciprocal lattice points we are interested in for this problem are

$$\Gamma \ : \quad k_x = 0 \quad k_y = 0 \quad k_z = 0 \qquad (\text{II.3.87})$$

$$X \ : \quad k_x = \frac{\pi}{a} \quad k_y = 0 \quad k_z = 0 \qquad (\text{II.3.88})$$

$$R \ : \quad k_x = \frac{\pi}{a} \quad k_y = \frac{\pi}{a} \quad k_z = \frac{\pi}{a} \ . \qquad (\text{II.3.89})$$

Therefore we obtain

– $\Gamma$ point:

$$M = \frac{\hbar^2}{E_1 a^2} \begin{pmatrix} 1 & 0 & 0 \\ 0 & 1 & 0 \\ 0 & 0 & 1 \end{pmatrix} \qquad \text{(II.3.90)}$$

– $X$ point:

$$M = \frac{\hbar^2}{E_1 a^2} \begin{pmatrix} -1 & 0 & 0 \\ 0 & 1 & 0 \\ 0 & 0 & 1 \end{pmatrix} \qquad \text{(II.3.91)}$$

– $R$ point:

$$M = \frac{\hbar^2}{E_1 a^2} \begin{pmatrix} -1 & 0 & 0 \\ 0 & -1 & 0 \\ 0 & 0 & -1 \end{pmatrix} . \qquad \text{(II.3.92)}$$

What are these effective mass tensors good for? In most cases the electrons are confined to energies close to the Fermi energy, and therefore the electron motion is along a constant energy surface. For nearly empty or nearly full bands the Fermi surface stays very close to the the $\Gamma$ or $R$ points, respectively. As long as the electrons will remain close to the maximum or minimum of the energy surface, the effective mass can be used to calculate physical properties, as in Eqs. I.3.19 and I.3.20. In particular, this is the case for semiconductors, where the electron and hole states are always close to the bottom and top of the bands.[3]

On the other hand, the effective mass tensor at the $X$ point is relevant only if the band filling has an intermediate value. But in the neighborhood of the $X$ point the constant energy cuts extend all over the Brillouin zone (similar to the neighborhood of the $X$ point in the two-dimensional half-filled band, illustrated in Figure II.3.13). Consequently, an electron can always move away from the $X$ point, and the effective mass will vary along its orbit. This makes the effective mass concept useless for arbitrary band filling. The point is well illustrated by taking the cyclotron mass, Eq. I.3.20, and selecting the magnetic field direction such that $M_{zz} < 0$. The formula then yields an imaginary $m^*$ whereas cyclotron orbits do exist at any band filling.

## 3.15 Hint: Cyclotron Frequency

Calculate the area of the electron's orbit in $k$ space as a function of energy. For low band fillings the constant energy cuts are ellipsoids; at fixed $k_z$ one obtains an ellipse in the $k_x$, $k_y$ plane. Use Eq. I.3.14 to calculate the cyclotron frequency.

---

[3] A simple formula for calculating the effective mass in semiconductors is given by Yu and Cardona [6] p. 65.

## 3.16 Solution: de Haas–van Alphen

The period of oscillation, $\Delta$, is related to the area of the electrons' orbits at that Fermi level (see Eq. I.3.15):

$$\Delta = \frac{1}{A}\frac{2\pi e}{\hbar c} = \frac{2\pi^2}{A\phi_0}, \tag{II.3.93}$$

where $\phi_0 = hc/2e = 2 \times 10^{-7}$ G/cm$^2$ is the flux quantum.

The area of the Fermi surface orbit is $A = 2\pi^2/\Delta\phi_0 = 1.61 \times 10^{15}$ cm$^{-2}$. The area of the first Brillouin zone is $A_{\mathrm{BZ}} = (\frac{2\pi}{a})^2 = 3.22 \times 10^{16}$ cm$^{-2}$. This is simply twice of the area of the orbit. Therefore the band is half-filled, and the electron density is $n = (1/2)(1/a^2) = 4.1 \times 10^{14}$ cm$^{-2}$.

## 3.17 Hint: Fermi Energy

For electrons, the allowed states in the Brillouin zone will be similar to the allowed phonon modes. We know from Problem 2.14 that the density of allowed states in the reciprocal space is given by

$$\text{Density} = \frac{Na^2 \sin 60^\circ}{4\pi^2}. \tag{II.3.94}$$

The Fermi wavenumber $k_F$ is defined such that the total number of electrons, $N$, fit within the Fermi sphere:

$$2k_F^2\pi\frac{Na^2 \sin 60^\circ}{4\pi^2} = N, \tag{II.3.95}$$

where the factor of 2 in front stands for the two spin states. From this we can solve for $k_F$ ($k_F = \sqrt{\frac{4\pi}{\sqrt{3}}\frac{1}{a}}$) and then for the Fermi energy,

$$E_F = \frac{\hbar^2 k_F^2}{2m} = 3.62\frac{\hbar^2}{ma^2}. \tag{II.3.96}$$

Using the given value of $a = 3$ Å, we obtain $E_F = 3.06$ eV.

# 4 Density of States

## 4.1 Solution: Density of States

According to Eq. I.4.5, the density of states can be calculated as a surface integral in the reciprocal space. For $E \sim | \boldsymbol{k} |^n$ the integrand is $k^{-(n-1)}$. (For massless particles $n = 1$; for massive particles $n = 2$.) In $d$ dimensions the surface area is $k^{d-1}$. This leads to $g \sim k^{d+n} \sim E^{(d-n)/n}$, in agreement with Table I.4.1.

## 4.2 Solution: Two-Dimensional Density of States

If $E(k)$ is a "smooth" function it can be expanded as

$$E(k) = E_0 + \frac{1}{2}\frac{\partial^2 E}{\partial k_i \partial k_j}(k_i - k_i^{(0)})(k_j - k_j^{(0)}) \qquad (\text{II.4.1})$$

in the vicinity of $\boldsymbol{k}^{(0)}$ near an extremum of $E(k)$.

Let us introduce a set of new variables as the linear combination of the old variables,

$$k_i^* = \sum_j \mathsf{a}_{ij}(k_i - k_i^{(0)}) \ . \qquad (\text{II.4.2})$$

We can always choose $\mathsf{a}_{ij}$ such that

$$\frac{\partial^2 E}{\partial k_i^* \partial k_j^*} = \delta_{ij} A \ . \qquad (\text{II.4.3})$$

In this case we will have

$$E = E_0 + \frac{1}{2}A(k_x^{*2} + k_y^{*2}) \ . \qquad (\text{II.4.4})$$

To evaluate the density of states, let us use the definition in Eq. I.4.2. The introduction of the new variable $k^*$ leads to

$$g(E) = \int_{BZ} \frac{\mathrm{d}^2 k}{(2\pi)^2}\delta[E - E(k)] = \frac{1}{\det|A|}\int_{BZ'} \frac{\mathrm{d}^2 k^*}{(2\pi)^2}\delta[E - \hbar^2 k^{*2}] \ . \qquad (\text{II.4.5})$$

161

This transformation turns the integral into the density of states of a two dimensional band with mass $m = 2$, $g_0 = 1/(4\pi)$. Therefore

$$g(E) = \frac{1}{\det|A|} \frac{1}{4\pi} \, . \tag{II.4.6}$$

Including the electron spin results in another factor of 2. The density of state is zero for energies $E < E_0$ and it jumps to a finite value for $E > E_0$.

Equation II.4.6 suggests that the density of states is constant for $E > E_0$ and this result is often quoted as a general property of the two-dimensional electron gas. However, we have to keep in mind that the result was obtained by neglecting the higher order terms in the expansion of the energy, Eq. II.4.1. The next nonvanishing term is usually fourth order in $k$ and this term yields a significant finite slope in $g(E)$. Numerical integration for a nonparabolic band clearly demonstrates this point (see Problem 4.3).

## 4.3 Solution: Two-Dimensional Tight-Binding

The energy dispersion specified in this problem leads to a "saddle point" in the $E(k)$ surface at $k = (\pi/a, 0)$ and symmetry equivalent points. At these points the energy is zero. It is very likely that the singularity seen in Figure I.4.1 is due to the saddle points in the energy surface.

Let us consider the neighborhood of the $k = (\pi/a, 0)$ point. Expansion of the cosine function gives $E = E_0[-\frac{1}{2}(k_x a - \pi)^2 + \frac{1}{2}(k_y a)^2]/4$ .

The density of states is

$$g(E) = \oint d\ell \frac{1}{|\partial E/\partial k|} \sim 8 \int_{\ell_1}^{\ell_2} d\ell \frac{4a}{E_0 a \sqrt{(k_x a - \pi)^2 + (k_y a)^2}}, \tag{II.4.7}$$

where the integration is along a constant energy line. The limits of integration are indicated in Figure II.4.1. In choosing the limits we used the symmetries of the integrand.

Let us look at energies slightly below zero and introduce the dimensionless parameters $\xi = \pi - k_x a$, $\eta = k_y a$, and $\varepsilon = -4E/E_0$; with these, $\varepsilon = 1/2\xi^2 - 1/2\eta^2$. Note that $(d\ell)^2 = (dk_x)^2 + (dk_y)^2 = a[(d\xi)^2 + (d\eta)^2]$ and, according to the constant energy condition, $d\varepsilon = \xi d\xi - \eta d\eta = 0$. The integrand becomes $d\ell/\sqrt{(k_x a - \pi)^2 + (k_y a)^2} = a d\xi/\eta = a d\xi/\sqrt{\xi^2 - 2\varepsilon}$. In terms of $\xi$, the limits of integration are $\xi_1 = \sqrt{2\varepsilon}$ and $\xi_2 \approx \pi/2$. (Of course, at this large value of $\xi$ the quadratic expansion of the cosine function is not valid. However, the integrand is relatively small in this regime and does not contribute much to the integral.) Finally, we obtain

$$g(E) = 32/a^2 E_0 \int_{\xi_1}^{\xi_2} \frac{d\xi}{\sqrt{\xi^2 - 2\varepsilon}} \sim \ln(\pi/2\sqrt{2\varepsilon}) \sim -\ln E \tag{II.4.8}$$

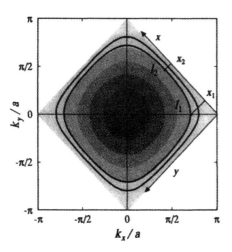

**Fig. II.4.1.** Integration path for calculating the density of states.

The density of states has a logarithmic divergence at $E = 0$.

Another solution to this problem starts with the original definition of the density of states: $g(E)dE$ is the number of states between $E$ and $E + dE$. Let us calculate this number simply by integrating the area between two constant energy lines in $\boldsymbol{k}$-space! With new wavenumber variables $x = k_y + (k_x - \pi/a)$ and $y = k_y - (k_x - \pi/a)$, the energy is $\varepsilon = xy/2$. The area under the $x = \varepsilon/2y$ hyperbola is $A = 1/2\varepsilon \ln(x_2/x_1)$, where $x_1 \sim 1/E$ and $x_2 \sim 1$ are the limits of integration (see Figure II.4.1). The density of states is $g(E) = dA/dE \sim \frac{d}{d\varepsilon}\varepsilon \ln 1/\varepsilon \sim -\ln\varepsilon$, as we obtained before.

## 4.4 Solution: Quasi-One-Dimensional Metal

The energy surface at $q_z = \frac{\pi}{2a}$ and the constant energy cuts are shown in Figure II.4.2, using Eq. I.4.8 with numerical values of $A = 1E_0$, $B = 0.1E_0$, and $b = 2a$.

For $d$ dimensions, the density of states is given by

$$g(E) = \int \frac{d\mathcal{S}}{(2\pi)^d} \frac{1}{|\boldsymbol{\nabla}E|} \cdot 2, \qquad (\text{II.4.9})$$

where the integral is over a constant energy surface (or $\boldsymbol{k}$-space) and the 2 is for spin. For three dimensions and one dimension, this formula reproduces Eqs. I.4.4 and I.4.5, respectively.

$\boldsymbol{\nabla}E$ is calculated from Eq. I.4.8 to be

$$\boldsymbol{\nabla}E = \begin{pmatrix} \frac{aA}{2}\sin q_x a \\ \frac{bB}{2}\sin q_y b \\ \frac{bB}{2}\sin q_z b \end{pmatrix} \qquad (\text{II.4.10})$$

for three dimensions. We then calculate $|\boldsymbol{\nabla}E|$,

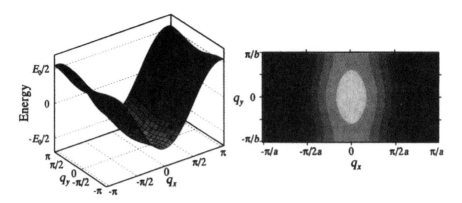

**Fig. II.4.2.** Quasi-one-dimensional band energy surface (right) and constant energy cuts (left). The third component of the wavenumber is fixed at $q_z = \frac{\pi}{2a}$.

$$|\nabla E| = \sqrt{\left(\frac{aA}{2}\right)^2 \sin^2 q_x a + \left(\frac{bB}{2}\right)^2 (\sin^2 q_y b + \sin^2 q_z b)} \ . \qquad \text{(II.4.11)}$$

Note that around half-filling $q_x$ varies very little, meaning that $\sin^2 q_x a$ does not become small. On the other hand, $B \ll A$ and therefore the approximation

$$|\nabla E| \approx \frac{aA}{2}|\sin q_x a| \qquad \text{(II.4.12)}$$

is always valid. We thus obtain a density of states of

$$g = \int \frac{\mathrm{d}\mathcal{S}}{4\pi^3} \frac{2}{aA} \frac{1}{|\sin q_x a|} \ . \qquad \text{(II.4.13)}$$

The right part of Figure II.4.2 shows that for a half-filled band most of the Fermi lines are nearly straight, extending from $-\pi/b$ to $\pi/b$. The surface integral in Eq. II.4.13 can then be approximated as:

$$g \approx \frac{1}{4\pi^3} \frac{2}{aA} \left( \int_{-\pi/b}^{\pi/b} \int_{-\pi/b}^{\pi/b} \mathrm{d}q_x \mathrm{d}q_y \right) \frac{1}{|\sin q_x a|} \qquad \text{(II.4.14)}$$

$$= \frac{2}{4\pi^3} \frac{2}{aA} \left(\frac{2\pi}{b}\right)^2 \frac{1}{|\sin q_x a|} \qquad \text{(II.4.15)}$$

$$= \frac{4}{\pi} \frac{1}{ab^2 A} \frac{1}{|\sin q_x a|} \ . \qquad \text{(II.4.16)}$$

Carrying out the same calculation for a truly one-dimensional metal, we start with the simplified equation for $\nabla E$:

$$\frac{\mathrm{d}\varepsilon}{\mathrm{d}q_x} = \frac{aA}{2} \sin q_x a \ . \qquad \text{(II.4.17)}$$

We then calculate the density of states:

$$g = \int \frac{dS}{2\pi} \frac{2}{aA} \frac{1}{|\sin q_x a|} \tag{II.4.18}$$

$$= \frac{4}{aA\pi} \frac{1}{|\sin q_x a|} . \tag{II.4.19}$$

In one dimension the integral was just the sum over the two points at $\pm q$, yielding a value of 2. Assuming each one-dimensional chain has a cross section of $b^2$, the density of states (per unit volume and unit energy) becomes

$$g = \frac{4}{aA\pi} \frac{1}{b^2} \frac{1}{|\sin q_x a|}. \tag{II.4.20}$$

What we have shown is that, as long as the energy is in the middle of the band, the quasi-one-dimensional result given by Eq. II.4.16 is equivalent to the real one-dimensional density of states in Eq. II.4.20.

## 4.5 Solution: Crossover to Quasi-One-Dimensional Metal

(a) We will use the results of Solution 4.4. In terms of the dimensionless variable $\varepsilon = E/E_0$ the band extends from $-\frac{1}{2}A - B$ to $\frac{1}{2}A + B$. The variable $\varepsilon'$ measures the energy from the bottom of the band; $\varepsilon' = \varepsilon + \frac{1}{2}A + B$.

For low energies $|q_x| \ll \pi/a$, as illustrated by the right part of Figure II.4.2. We expand the energy given in Eq. I.4.9 as

$$\varepsilon = -\frac{A}{2} + \frac{A}{4}a^2 q_x^2 - \frac{B}{2}(\cos q_y b + \cos q_z b)$$

$$\text{or } \varepsilon' = \frac{A}{4}a^2 q_x^2 + B - \frac{B}{2}(\cos q_y b + \cos q_z b) . \tag{II.4.21}$$

At low energies the constant energy cuts look like pancakes: Extended in the $q_y$ and $q_z$ directions, but short in the $q_x$ direction.

The crossover from 1D to 3D behavior can be defined in various ways. For example, one may consider the energy when the density of states begins to differ significantly from the 1D density of states (see Figure I.4.2), or look for the energy when the Fermi surface becomes open. An inspection of Figure II.4.2 suggests that the corresponding energy will be in the range of $E_0 B$; inserting $q_x = 0$ and $q_y = q_z = \pi/b$ into Eq. II.4.21 results in $\varepsilon' = 2B$.

The volume inside the Fermi surface, $V_{FS}$, tells us the crossover electron density: $n_{cross} = (2/8\pi^3)V_{FS}$. The shape of the Fermi surface is not simple; at best we can estimate the volume. First we determine the thickness of the "pancake", $\delta q_x$, by inserting $\varepsilon' = 2B$ into Eq. II.4.21 for $q_y = q_z = 0$:

$$\delta q_x = \frac{2}{a}\sqrt{\frac{2B}{A}} . \tag{II.4.22}$$

We approximate the Fermi surface by two planes extending from $-\pi/b$ to $\pi/b$ in $q_y$ and $q_z$, and separated by a distance $\delta q_x$ in the $q_x$ direction. The volume within these two planes is then

$$V_{FS} = \left(\frac{2\pi}{b}\right)^2 \delta q_x = \frac{8\pi^2}{ab^2}\sqrt{\frac{B}{A}} \ . \tag{II.4.23}$$

The "crossover" density is now calculated:

$$n_{\text{cross}} = \frac{2}{(2\pi)^3} V_{FS} = \frac{2}{\pi}\frac{1}{ab^2}\sqrt{\frac{B}{A}} \ . \tag{II.4.24}$$

The factor $\frac{2}{\pi}$ is of order unity and can be dropped.

(b) For very low values of the density of states, we can expand the energy around the bottom of the band:

$$\varepsilon' = \frac{1}{4}\left[Aa^2q_x^2 + Bb^2(q_y^2 + q_z^2)\right] \ . \tag{II.4.25}$$

For bands with a parabolic bottom, the density of states can be calculated as follows. By definition, $g(E)$ is the function which turns the momentum integral into an energy integral,

$$I = \int \frac{2}{(2\pi)^3} F(E)\mathrm{d}q_x\,\mathrm{d}q_y\,\mathrm{d}q_z = \int F(E)g(E)\mathrm{d}E \ , \tag{II.4.26}$$

where $F(E)$ is an arbitrary function. We introduce the new dimensionless variables

$$q_x' = \sqrt{\frac{Aa^2}{4}}q_x, \quad q_y' = \sqrt{\frac{Bb^2}{4}}q_y, \quad q_z' = \sqrt{\frac{Bb^2}{4}}q_z \tag{II.4.27}$$

and with these variables the energy is

$$\varepsilon' = q_x'^2 + q_y'^2 + q_z'^2 \equiv \mathbf{q}'^2 \ . \tag{II.4.28}$$

We rewrite the integral in Eq. II.4.26 as

$$I = \int \frac{2}{(2\pi)^3}\sqrt{\frac{4}{Aa^2}\frac{4}{Bb^2}}F(E)\mathrm{d}^3\mathbf{q}' \ , \tag{II.4.29}$$

where $\mathrm{d}^3\mathbf{q}' = 4\pi q'^2\mathrm{d}q' = 4\pi\varepsilon'\frac{1}{2\sqrt{\varepsilon'}}\mathrm{d}\varepsilon'$. We therefore obtain a density of states

$$g(\varepsilon') = \frac{16}{\pi^2}\frac{1}{ab^2}\frac{1}{\sqrt{AB}}\sqrt{\varepsilon'} \ . \tag{II.4.30}$$

The result is in agreement with Table I.4.1.

A similar, but much simpler calculation yields the density of states for a truly 1D system for $\varepsilon' \ll 1$:

$$g(\varepsilon') = \frac{2}{\pi}\frac{1}{ab^2}\frac{1}{\sqrt{A\varepsilon'}} \ . \tag{II.4.31}$$

In the one-dimensional case the density of states diverges at low band filling, whereas in three-dimensions it approaches zero.

## 4.6 Solution: Phonon Mode of Two-Dimensional System

We chose $\omega_0 = 1$, $a = 1$, and we took a grid of $63 \times 63$ points from $k_x = 0$ to $k_x = \pi$ and $k_y = 0$ to $k_y = \pi$. We used a worksheet program on a personal computer to calculate the density of states and sum how many $k$-states are within a given energy interval. The resultant histogram is shown in Figure II.4.3.

The Debye density of states is obtained from the small wavenumber expansion,

$$\omega = \frac{\omega_0}{\sqrt{2}} a|k| \ . \tag{II.4.32}$$

For a single band in two dimensions, Eq. I.4.5 becomes

$$g = \int \frac{d\ell}{4\pi^2} \frac{1}{|v|} \ , \tag{II.4.33}$$

where $d\ell = \sqrt{(dk_x)^2 + (dk_y)^2}$ is along a constant frequency line on the $k$ plane.[1] The velocity is $|v| = a\omega_0/\sqrt{2}$ and in polar coordinates $dS = k\,d\phi$.

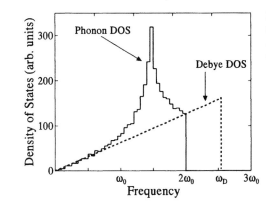

**Fig. II.4.3.** Phonon density of states in two dimensions *vs.* energy. Also shown is the appropriate Debye density of states.

Therefore

$$g_D = \frac{1}{4\pi^2} \int \frac{k\,d\phi}{a\omega_0/\sqrt{2}} = \frac{1}{2\pi} \frac{2}{a^2\omega_0^2} \omega \ . \tag{II.4.34}$$

The Debye frequency is determined by the condition

$$\int_0^{\omega_D} g_D(\omega)\,d\omega = \int g(\omega)\,d\omega \ , \tag{II.4.35}$$

where $g(\omega)$ is the "true" density of states. This results in $\omega_D = \sqrt{2\pi}\omega_0$, as indicated in the figure.

There are several interesting features to point out:

---

[1] We dropped the $\hbar$ factor in Eq. I.4.5 and defined $g$ as states per unit frequency, instead of states per unit energy.

- The Debye and "true" density of states (DOS) coincide at low energies.
- At $\omega = 2\omega_0$ there is another coincidence between the Debye and "true" DOS; this is a unique feature of this particular system.
- The Debye frequency is *larger* than the upper cutoff frequency of the real system.
- The van Hove singularity seen in Figure II.4.3 at $\omega = \sqrt{2}\omega_0$ is due to the saddle points in the energy surface shown is Figure I.4.3.
- Although at the highest phonon frequencies ($\omega = 2\omega_0$) the group velocity is zero, $|v| = 0$, the density of state does not diverge.

## 4.7 Solution: Saddle Point

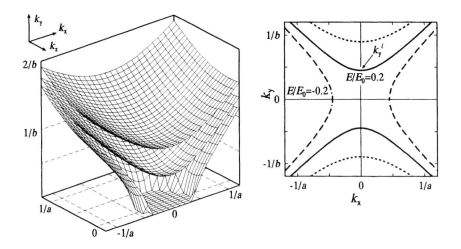

**Fig. II.4.4.** Constant energy cuts of the energy dispersion described by Eq. I.4.11. The energies are $E/E_0 = -0.2$, 0.2, and 0.8.

The parameters $E^*$, $a$, $b$, and $c$ characterize the band; note the negative sign of the $b$ term. Without losing the generality of the solution, we can take $a = c$ in Eq. I.4.11. If the energy happens to be anisotropic in the $k_x, k_z$ plane, a simple coordinate transformation would bring it to an isotropic form.[2]

A few of the constant energy cuts are shown in Figure II.4.4. Note the change in the topology of these surfaces as the energy crosses from $E > E_0$ to $E < E_0$.

We would like to calculate the density of states,

---

[2] A similar problem for a two-dimensional system is Problem 4.3.

$$g(E) = \frac{1}{4\pi^3} \int \frac{dS}{|\nabla E|} , \qquad \text{(II.4.36)}$$

where $\nabla E \equiv \partial E/\partial \mathbf{k}$. Let us first set $k_z = 0$! In the $k_z = 0$ plane we can solve for $\nabla E$ as follows:

$$k_y = \sqrt{\frac{a^2}{b^2}k_x^2 - \frac{E - E_0}{E^* b^2}} \qquad \text{(II.4.37)}$$

$$dk_y = \frac{a^2}{b^2}\frac{k_x}{k_y}dk_x \qquad \text{(II.4.38)}$$

$$|\nabla E| = 2E^* b^2 \sqrt{\left(\frac{a^4}{b^4}\right)^2 k_x^2 + k_y^2} . \qquad \text{(II.4.39)}$$

Using the rotational symmetry about the $y$-axis, we get $dS = 2\pi k_x d\ell$ where $d\ell = \sqrt{dk_x^2 + dk_y^2}$. The density of states then becomes (remembering a factor of 2 for the two branches)

$$g(E) = \frac{1}{4\pi^3} 2 \int \frac{2\pi k_x dk_x \sqrt{1 + \frac{a^4}{b^4}\left(\frac{k_x}{k_y}\right)^2}}{2E^* b^2 \sqrt{\frac{a^4}{b^4}k_x^2 + k_y^2}} \qquad \text{(II.4.40)}$$

$$= \frac{1}{2\pi^2 E^* b^2} \int \frac{k_x dk_x}{k_y} , \qquad \text{(II.4.41)}$$

and the integration is along the positive $k_x$ axis. But, according to Eq. II.4.38, we have

$$\frac{k_x}{k_y} = \frac{b^2}{a^2}\frac{dk_y}{dk_x} , \qquad \text{(II.4.42)}$$

and therefore by using Eq. I.4.11 we obtain

$$g(E) = \frac{1}{2\pi^2 E^* a^2} k_y \Big|_{k_y^l}^{k_y^h} = \frac{1}{2\pi^2 E^* a^2} \sqrt{\frac{a^2}{b^2}k_x^2 - \frac{E - E_0}{E^* b^2}}\Bigg|_{k_y^l}^{k_y^h} , \qquad \text{(II.4.43)}$$

where $k_y^l$ and $k_y^h$ are the limits of integration.

The right panel in Figure II.4.4 illustrates the constant energy cuts in the $k_x, k_y$ plane. The lower limit of the integration is zero for $E < E_0$ and is $k_y^l = \frac{1}{a}\sqrt{(|E - E_0|)/E^*}$ for $E > E_0$. The upper limit, $k_y^h$, is on the order of magnitude of $k_y^h = \frac{\pi}{a} \gg \frac{1}{a}\sqrt{(|E - E_0|)/E^*}$. The energy dependence in the neighborhood of the upper limit is negligible:

$$g(E) = \begin{cases} \frac{1}{2\pi^2 ab}k_y^h - \frac{1}{2\pi^2 a^2 bE^*}\sqrt{\frac{E_0 - E}{E^*}} & \text{for } E < E_0 , \\ \frac{1}{2\pi^2 ab}k_y^h & \text{for } E > E_0 . \end{cases} \qquad \text{(II.4.44)}$$

This density of states is plotted in Figure II.4.5.

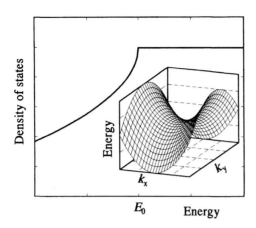

$E_0$    Energy

Fig. II.4.5. The density of states as a function of energy showing its functional change as the energy becomes larger than $E_0$. The inset illustrates the saddle point of the energy dispersion.

## 4.8 Solution: Density of States in Superconductors

We start with the definition of the density of states and a few formal operations: The number of states between $\epsilon$ and $\epsilon + d\epsilon$ is

$$g(\epsilon)d\epsilon = g(\epsilon)\frac{d\epsilon}{dE}dE = g'(E)dE \; , \qquad (\text{II}.4.45)$$

which is the number of states between $E$ and $E + dE$.

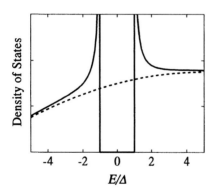

Fig. II.4.6. Density of single-electron states in a superconductor (solid line) Also plotted is the density of states for $\Delta = 0$ (dashed line).

For constant ($\boldsymbol{k}$-independent) $\Delta$, the function $E(\epsilon)$ is well-defined. The result is

$$g'(E) = g(\epsilon(E))\frac{|E|}{\sqrt{E^2 - \Delta^2}} \; , \qquad (\text{II}.4.46)$$

as plotted in Figure II.4.6. The "pile up" of states at $E = \pm\Delta$ compensates for the states missing around $E = 0$, and keeps total number of states, i.e., the area under the curve, constant. We note that this single result is not valid if $\Delta$ is $\boldsymbol{k}$-dependent.

## 4.9 Solution: Energy Gap

Equation I.3.7 and the solution to Problem 3.8 can be used here for the band structure. From Eq. I.4.6 one obtains the density of states

$$g(E) = g_0 \frac{|E - E_F|}{\sqrt{(E - E_F)^2 - \Delta^2}} , \qquad \text{(II.4.47)}$$

where $\Delta = U_0/2$. As long as the limits of integration are much larger than $\Delta$, this function satisfies the requirement that the total number of states, $\int g(\mathcal{E})d\mathcal{E}$, does not depend on the gap. This result is similar to the density of states obtained in Solution 4.8 for a superconductor.

## 4.10 Solution: Density of States for Hybridized Bands

According the the definition of the DOS we obtain (similar to Problem 4.8)

$$g(E') = g_c(E_c) \frac{dE_c}{dE'} . \qquad \text{(II.4.48)}$$

By inverting the expression, we get

$$E' = (E_c + E_0)/2 \pm \sqrt{(E_c - E_0)^2/4 + \Delta^2} , \qquad \text{(II.4.49)}$$

and $E_c = E' - E_0 - \Delta^2/E'$. Accordingly, $dE_c/dE' = 1 + \left(\frac{\Delta}{E'}\right)^2$. With a constant DOS for the conduction band $E_c$, we obtain

$$g(E') = g_0 \left[1 + \left(\frac{\Delta}{E'}\right)\right]^2 . \qquad \text{(II.4.50)}$$

At this point we also have to specify what the bandwidth is and where the density of states becomes zero. Using Eq. II.4.49 the energies $E_1$ and $E_2$ are converted to

$$E'_1 = (E_1 + E_0)/2 - \sqrt{(E_1 - E_0)^2/4 + \Delta^2} \qquad \text{(II.4.51)}$$
$$E'_2 = (E_2 + E_0)/2 + \sqrt{(E_2 - E_0)^2/4 + \Delta^2} . \qquad \text{(II.4.52)}$$

This is, however, a minor change, as in most cases $\Delta \ll W$. In good approximation $E'_1 \sim E_1$ and $E'_2 \sim E_2$. On the other hand, the new DOS is also zero in a narrow energy range around $E_0$. This is due to the fact that for $E'$ close to $E_0$ the expression $E_c = E' - \Delta^2/E'$ leads to values outside of the original bandwidth, and for those values $g(E_c)$ is zero. The corresponding energies are

$$E'_3 = (E_2 + E_0)/2 - \sqrt{(E_2 - E_0)^2/4 + \Delta^2} \qquad \text{(II.4.53)}$$
$$E'_4 = (E_1 + E_0)/2 + \sqrt{(E_1 - E_0)^2/4 + \Delta^2} . \qquad \text{(II.4.54)}$$

According to these considerations, our original broad band splits into two bands. The lower band ranges from $E'_1$ to $E'_3$, the upper one is between $E'_4$ and $E'_2$ in the $\Delta \ll W$ limit. And if $E_0$ is not too close to the band edges, the DOS is nonzero from $E_1$ to $E_0 - 2\Delta^2 / (E_2 - E_0)^2$ and from $E_0 + 2\Delta^2 / (E_1 - E_0)^2$ to $E_2$. This function is plotted in Figure II.4.7.

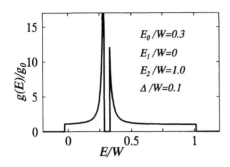

**Fig. II.4.7.** Normalized density of states of two, hybridized bands. The parameters of the calculation are indicated in the figure.

The small range of zero DOS around $E' = E_0$ is very important from the point of view of the total number of states in the band. The function described by Eq. II.4.50 is strongly divergent at $E' = E_0$; without the cutoff, the integral $N = \int g(E)\, dE$ would diverge. It is easy to show that with the limits determined here, the total number of states is $N = 2W g_0$. This means that no states were created or destroyed by the hybridization. On the other hand, it is a little counterintuitive to see that the DOS becomes zero right where the original DOS had a Dirac-delta peak. This artifact is the consequence of the overly simplified form we used to create the new energies from the old ones, Eq. II.4.49. In particular, a more realistic formula could mix the original conduction band states belonging to different $k$-values into the new band states. This would lead to the elimination of the unphysical zero DOS around $E' = E_0$ seen in Figure II.4.7. Furthermore, any overlap between the atomic orbitals involved in the narrow band would lead to a finite bandwidth; this effect also eliminates the "hole" in the DOS.

## 4.11 Solution: Infinite-Dimensional DOS

Let us consider the calculation of the DOS by the Monte Carlo method: We choose a random value of the $d$-dimensional $k = (k_1, k_2, \ldots, k_i)$ and calculate the corresponding energy $E(k) = \frac{E_0}{2d} \sum_1^d (- \cos k_i a + 1)$, where $E_0$ is the bandwidth. To determine the DOS we count how many events result in an energy between $E$ and $E + dE$. If viewed this way, the calculation becomes equivalent to finding the distribution of a random number generated by the sum of $d$ independent random numbers, ranging from zero to $E_0$. We can use the central limit distribution theorem to obtain the answer. The distribution is centered around $\overline{E} = dE_0/2$ with a Gaussian shape:

$$g(E) = \frac{1}{E_0\sqrt{\pi d}} e^{-(E-dE_0/2)^2/E_0^2 d} = \frac{1}{E_0\sqrt{\pi/d}} e^{-[(E/E_0-1/2)\sqrt{d}]^2} \quad \text{(II.4.55)}$$

With $E_0 = 1$ this corresponds to a sharp peak of width $\sim 1/\sqrt{d}$, centered at energy $= 1/2$ in Figure I.4.4.

Using this line of argument, Metzner and Vollhardt[3] concluded that the only meaningful way to generalize the tight-binding model to $d \to \infty$ is by scaling the bandwidth parameter as $E_0 \propto \sqrt{d}$.

## 4.12 Solution: Two-Dimensional Electron Gas

The quasi-two-dimensional electron gas in a box of size $d$ will have discrete energy bands. At $k_x = k_y = 0$, the energies of these bands are given by $E_n = \frac{\hbar^2 k6@}{2m} \frac{\pi^2}{d^2} n^2$. A sketch of the $k_y = 0$ cut in the energy surface is displayed in Figure II.4.8.

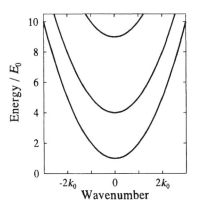

Fig. II.4.8. $k_y = 0$ cut through the energy bands for an electron gas confined to a potential well of width $d$. The energy scale is $E_0 = (\hbar^2/2)(\pi^2/d^2)$, the wavenumber scale is $k_0 = 2\pi/d$

We number the energy bands by the variable $n$. To calculate the density of states, let us first consider the $n = $ fixed case. From the two-dimensional version of Eq. I.4.4,

$$g = 2 \int \frac{dS}{(2\pi)^2} \frac{1}{|\partial E/\partial k|} = 2 \int_0^{2\pi} d\phi \frac{k}{(2\pi)^2} \frac{1}{k} \frac{m}{\hbar^2} = \frac{1}{\pi} \frac{m}{\hbar^2} \quad \text{(II.4.56)}$$

and

$$g = 0 \text{ for } E < \frac{\hbar^2}{2m} \frac{\pi^2}{d^2} n^2 , \quad \text{(II.4.57)}$$

where the 2 in front is for spin. Therefore the density of states for any band is a step function. As the energy $E$ increases, more bands will enter into the density of states. The total density of states as a function of $E$ is shown in Figure II.4.9.

---

[3] W. Metzner and D. Vollhardt, *Phys. Rev. Lett.*, **62**, 324 (1989).

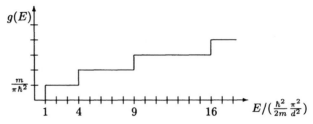

**Fig. II.4.9** Density of states for a quasi-two-dimensional free-electron gas.

For high energies (high relative to $\frac{\hbar^2}{2m}\frac{\pi^2}{d^2}$), the many steps in the density of states look like a parabolic function ($g \propto E^{1/2}$).

$$E = \frac{\hbar^2}{2m}\frac{\pi^2}{d^2}n^2 \tag{II.4.58}$$

$$g = \sum_{n=1}^{n} \frac{1}{\pi}\frac{m}{\hbar^2} = \frac{m}{\pi\hbar^2}n = \frac{m}{\pi\hbar^2}\frac{d}{\pi}\sqrt{\frac{2mE}{\hbar^2}} \tag{II.4.59}$$

$$g(E) = d\frac{1}{2\pi^2}\left(\frac{2m}{\hbar^2}\right)^{3/2}E^{1/2}. \tag{II.4.60}$$

We note that $g(E)/d$ is equivalent to the three-dimensional density of states. The electrons will behave two-dimensionally if $k_B T$ is less than the distance between steps in $g(E)$, which is approximately given by $\frac{\hbar^2}{2m}\frac{\pi^2}{d^2}(2^2 - 1^2)$. For $d = 100$ Å we obtain a temperature of

$$T < \frac{\hbar^2}{2m}\frac{\pi^2}{d^2}\frac{1}{k_B} = 130 \text{ K} \tag{II.4.61}$$

below which the system will act two-dimensionally. Of course for higher band fillings the separation may be larger. Also, in real systems we can only produce a finite potential well. This puts a lower limit on $d$, since the ground state must be a bound state in the $z$ direction, with a clear energy gap up to the excited states. To estimate $d$ for a potential of 100 meV, we use

$$\frac{\hbar^2}{2m}\frac{\pi^2}{d^2} < \frac{1}{2}V \tag{II.4.62}$$

to obtain $d > 20$ Å. Using Eq. II.4.61, the temperature limit of 20 mK results in size limit of $d < 7000$ Å. Hence for the "real" case specified, the electron gas must be confined within 20 Å $< d <$ 7000 Å for it to act two-dimensionally.

# 5 Elementary Excitations

## 5.1 Solution: Tight-Binding Model

(a) The Schrödinger equation, using the wavefunction $\phi(k, s)$ given in the hint, is

$$t\left[\sum_i \sum_s a^+_{i+1,s}a_{i,s} + a^+_{i,s}a_{i+1,s}\right] \frac{1}{\sqrt{N}} \sum_p e^{ikx_p} a^+_{p,s}|0\rangle$$

$$= E\frac{1}{\sqrt{N}} \sum_p e^{ikx_p} a^+_{p,s}|0\rangle \; . \qquad (\text{II}.5.1)$$

By using the commutation relation given in Eq. I.5.3, this can be rewritten as

$$\frac{t}{\sqrt{N}}\left[\sum_p e^{ik(p+1)a} a^+_{p+1,s}(1 - a^+_{p,s}a_{p,s})\right.$$

$$\left. + e^{ik(p-1)a} a^+_{p-1,s}(1 - a^+_{p,s}a_{p,s})\right]|0\rangle$$

$$= \frac{E}{\sqrt{N}} \sum_p e^{ikpa} a^+_{p,s}|0\rangle \; . \qquad (\text{II}.5.2)$$

The $a^+a^+a|0\rangle$ terms are zero. The equation is then simplified and can be solved as follows:

$$\frac{t}{\sqrt{N}}\left(e^{ika} + e^{-ika}\right) \sum_p e^{ikpa} = \frac{E}{\sqrt{N}} \sum_p e^{ikpa} \qquad (\text{II}.5.3)$$

$$2t\left(\frac{e^{ika} + e^{-ika}}{2}\right) = E \qquad (\text{II}.5.4)$$

or $E = 2t \cos ka$.

(b) We now carry out the same calculation as in part (a), only now we must use the two-electron wavefunction. The Schrödinger equation now becomes

$$t\left[\sum_i\sum_s a_{i+1,s}^+ a_{i,s} + a_{i,s}^+ a_{i+1,s}\right]\frac{1}{N}\sum_{p,q}e^{ik x_p}e^{ik' x_q}a_{p,s}^+ a_{q,s'}^+|0\rangle$$

$$= E\frac{1}{N}\sum_p e^{ik x_p}e^{ik' x_q}a_{p,s}^+ a_{q,s'}^+|0\rangle . \qquad \text{(II.5.5)}$$

Using Eq. I.5.3, we obtain

$$\frac{t}{N}\left[\sum_{p,q}e^{ik(p+1)a}e^{ik'qa}a_{p+1,s}^+(1-a_{p,s}^+a_{p,s})a_{q,s'}^+\right.$$

$$\left. + e^{ik(p-1)a}e^{ik'qa}a_{p-1,s}^+(1-a_{p,s}^+a_{p,s})a_{q,s'}^+\right]|0\rangle$$

$$= \frac{E}{N}\sum_{p,q}e^{ikpa}e^{ik'qa}a_{p,s}^+ a_{q,s'}^+|0\rangle . \qquad \text{(II.5.6)}$$

And applying the commutation relation again, we get

$$\frac{t}{N}\left[\sum_{p,q}e^{ik(p+1)a}e^{ik'qa}a_{p+1,s}^+ a_{q,s'}^+ + e^{ikpa}e^{ik'(q+1)a}a_{p+1,s}^+ a_{q,s'}^+(1-a_{q,s}^+ a_{q,s})\right.$$

$$\left. + e^{ik(p-1)a}e^{ik'qa}a_{p-1,s}^+ a_{q,s'}^+ + e^{ikpa}e^{ik'(q-1)a}(1-a_{q,s}^+ a_{q,s})\right]|0\rangle$$

$$= \frac{E}{N}\sum_{p,q}e^{ikpa}e^{ik'qa}a_{p,s}^+ a_{q,s'}^+|0\rangle . \qquad \text{(II.5.7)}$$

This is then solved just like in part (a) to obtain the final solution:

$$E = 2t\cos ka + 2t\cos k'a . \qquad \text{(II.5.8)}$$

(c) When $k = k'$ and $s = s'$ (i.e., the two electrons are in the same state with the same spin), the wavefunction is

$$|\phi(k,k,s,s)\rangle = \frac{1}{N}\sum_{p,q}e^{ik(x_p+x_q)}a_{p,s}^+ a_{q,s}^+|0\rangle = 0 . \qquad \text{(II.5.9)}$$

This shows that the use of electron creation and annihilation operators ensures the validity of the Pauli exclusion principle.

(d) Since the electron energies add up, $E(k_{tot}) = E(k) + E(k')$. However, $k$ and $k'$ are not fixed; for a given $k_{tot}$, we have $k' = k_{tot} - k$. Therefore, for most of its range of arguments, $E(k_{tot})$ is not a single-valued function. The two-electron excitations have a spectrum of continuous energies, with a lower and upper bound, as shown in the Figure II.5.1. The bounds are given by $d\left[E(k) + E(k_{tot} - k)\right]/dk = 0$. From the condition $\sin ka = \sin(k_{tot} - k)a$ we obtain $k = k_{tot}/2$ or $k = \pi/a + k_{tot}/2$. Accordingly the lower and upper bounds satisfy

$$E_{low} = -4t\cos\frac{k_{tot}}{2}a \qquad \text{(II.5.10)}$$

$$E_{\text{high}} = 4t \cos \frac{k_{\text{tot}}}{2} a \ . \qquad \text{(II.5.11)}$$

The range between these two energy values is filled up with possible two-electron excitations.

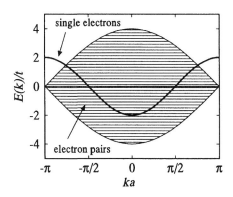

**Fig. II.5.1.** Single-electron and two-electron excitations in the tight-binding model.

## 5.2 Solution: Hybridization of Energy Bands

(a) We take $\Delta = 0$ and we first solve the problem with no $d$ band. The Hamiltonian for the conduction band has been investigated in Problem 5.1; the wavefunctions are

$$|k\rangle = \frac{1}{\sqrt{N}} \sum_j e^{ikR_j} c_j^+ |0\rangle \ , \qquad \text{(II.5.12)}$$

where $N$ is the number of sites, $R_j = ja$ is the position of the $j$th site, and $|0\rangle$ is the vacuum state. The energy is

$$E_c(k) = -2t \cos ka \ . \qquad \text{(II.5.13)}$$

According to the periodic boundary condition, the wavevector $k$ is quantized and there is altogether $N$ $k$ values corresponding to the $N$ quantum states of the spinless electron.

The impurity Hamiltonian is trivial: The wavefunctions are $d_i^+ |0\rangle$ with degenerate energies $E_0$. Due to the degeneracy, we are free to choose any linear combination of these wavefunctions as a basis set. In analogy to the delocalized electrons, we take

$$|q\rangle = \frac{1}{\sqrt{N}} \sum_l e^{iqR_l} d_l^+ |0\rangle \ . \qquad \text{(II.5.14)}$$

Again, the periodic boundary condition leads to $N$ states of different $q$ vectors.

The combined Hamiltonian with $\Delta = 0$ has $2N$ states and the matrix representation is diagonal in terms of the wavefunction described above. Accordingly, the energy spectrum has two contributions: An $E_c(k) = -2t\cos ka$ part (due to the $s$ electrons) and an $E_d(k) = E_0$ (due to the $d$ electrons), as illustrated by the dashed lines in Figure II.5.2.

(b) As we now turn to $\Delta \neq 0$ the matrix elements of the full Hamiltonian are

$$
H = \begin{bmatrix}
E_c(k_1) & 0 & \cdots & 0 & \Delta_{k_1}^{q_1} & \Delta_{k_1}^{q_2} & \cdots & \Delta_{k_1}^{q_N} \\
0 & E_c(k_2) & \cdots & 0 & \Delta_{k_2}^{q_1} & \Delta_{k_2}^{q_2} & \cdots & \Delta_{k_2}^{q_N} \\
\vdots & \vdots & \ddots & \vdots & \vdots & \vdots & \ddots & \vdots \\
0 & 0 & \cdots & E_c(k_N) & \Delta_{k_N}^{q_1} & \Delta_{k_N}^{q_2} & \cdots & \Delta_{k_N}^{q_N} \\
\Delta_{k_1}^{q_1*} & \Delta_{k_2}^{q_1*} & \cdots & \Delta_{k_N}^{q_1*} & E_0 & 0 & \cdots & 0 \\
\Delta_{k_1}^{q_2*} & \Delta_{k_2}^{q_2*} & \cdots & \Delta_{k_N}^{q_2*} & 0 & E_0 & \cdots & 0 \\
\vdots & \vdots & \ddots & \vdots & \vdots & \vdots & \ddots & \vdots \\
\Delta_{k_1}^{q_N*} & \Delta_{k_2}^{q_N*} & \cdots & \Delta_{k_N}^{q_N*} & 0 & 0 & \cdots & E_0
\end{bmatrix}
$$

$$\text{(II.5.15)}$$

where $\Delta_k^q = \Delta\langle 0|\frac{1}{N}\sum_{ilj} e^{-iqR_l}e^{ikR_j}c_j(c_i^+ d_i + c_i d_i^+)d_l^+|0\rangle$ and $\Delta^*$ is the complex conjugate.

It does not take much to show that $\Delta_k^q = \Delta\delta_{kq}$. This makes the calculation of the energy eigenvalues rather straightforward. The Hamiltonian matrix simplifies to

$$
H = \begin{bmatrix}
E_c(k_1) & 0 & 0 & \cdots & & 0 & 0 & \Delta_{k_1}^{q_N} \\
0 & E_c(k_2) & 0 & \cdots & & 0 & \Delta_{k_2}^{q_{N-1}} & 0 \\
0 & 0 & \ddots & & & & 0 & 0 \\
\vdots & \vdots & & E_c(k_N) & \Delta_{k_1}^{q_1} & & \vdots & \vdots \\
& & & \Delta_{k_N}^{q_1*} & E_0 & & & \\
0 & 0 & \cdot & & & \ddots & 0 & 0 \\
0 & \Delta_{k_2}^{q_{N-1}*} & 0 & \cdots & & 0 & E_0 & 0 \\
\Delta_{k_1}^{q_N*} & 0 & 0 & \cdots & & 0 & 0 & E_0
\end{bmatrix}
$$

$$\text{(II.5.16)}$$

By reordering the rows and columns $H$ can be block diagonalized into $2 \times 2$ blocks. The determinant condition $|H - E'| = 0$ leads to

$$
\begin{vmatrix}
E_c(k) - E' & \Delta \\
\Delta & E_0 - E'
\end{vmatrix} = 0 ,
$$

$$\text{(II.5.17)}$$

for each value of $k$. The result is

$$
E'(k) = (E_c + E_0)/2 \pm \sqrt{(E_c - E_0)^2/4 + \Delta^2} .
$$

$$\text{(II.5.18)}$$

For wavevectors when $| (E_c - E_0) | /t \gg \Delta$, this expression reduces to $E_1'(k) = E_c(k)$ and $E_2'(k) = E_0$. If the $d$ level lies outside of the conduction band, not much happens to the energies. However, if the conduction band energy becomes close to the $d$ level, in the neighborhood of $E(k) = E_d$, we see *hybridization*: The $d$ states and the conduction band states are strongly mixed and the energy levels "repel" each other. The resulting energy spectrum is shown in Figure II.5.2.

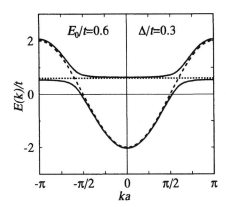

**Fig. II.5.2.** Energy dispersion of two hybridized bands. The dashed and dotted lines represent the original $s$ and $d$ bands, respectively.

A similar model can be used to describe impurity levels in a conduction band. Note the remote analogy between this problem and the phonon problem of the harmonic chain with a mass defect, Problem 2.8.

## 5.3 Solution: Polarons

The trial function suggested here, $u = u_0 \tanh x/a_0$, leads to an exactly solvable one-dimensional Schrödinger equation for the electron. In reality, the problem is much more complicated, since the lattice distortion may be different; it may even fluctuate in time. The selection of a simple lattice deformation function removes most of the complexity of the coupled electron–phonon problem. This treatment is a typical example of a "mean field" approximation, where the problem becomes solvable by cleverly eliminating all but a few degrees of freedom and looking for a self-consistency in the remaining ones.

The Schrödinger equation for the electron is

$$- \hbar/2m \; \mathrm{d}^2\psi/\mathrm{d}x^2 - V_0 \cosh^2 x/a_0 \; \psi = E\psi \; , \qquad (\text{II.5.19})$$

where $V_0 = \lambda u_0/a_0$ characterizes the strength of the interaction. According to textbooks (*e.g.*, Landau and Lifshitz [9] p. 72), the equation can be solved by introducing the new variable, $r = \tanh x/a_0$, and the wavefunctions can be expressed in terms of Legendre polynomials. The energy spectrum is

$$E_{\text{electron}} = \frac{-\hbar^2}{8ma_0^2}\left[-(1+2n)+\sqrt{1+8ma_0^2V_0/\hbar^2}\right]^2, \qquad \text{(II.5.20)}$$

with $n = 0, 1, 2, \ldots$. The ground-state energy is

$$E_0 = -\frac{\hbar^2}{8ma_0^2}\left[\sqrt{1+8ma_0^2V_0/\hbar^2}-1\right]^2. \qquad \text{(II.5.21)}$$

For attractive electron–phonon interactions (i.e., $\lambda > 0$) the ground state has negative energy; the electron gains energy by creating the lattice distortion.

The elastic energy cost of the distortion will be

$$E_{\text{lattice}} = 1/2\int B(\frac{du}{dx})^2\ dx = \frac{1}{2}\frac{Bu_0^2}{a_0}\int\frac{dz}{\cosh^4 z} = \frac{1}{3}Bu_0^2/a_0. \qquad \text{(II.5.22)}$$

The maximum energy gain from the coupled electron–phonon system can be found by minimizing the total energy $E_{\text{total}} = E_{\text{electron}} + E_{\text{lattice}}$ with respect to the two variables $u_0$ and $a_0$. It is helpful to introduce the $\xi$ and $\eta$ dimensionless variables by $u_0 = 3\lambda\eta/2B$ and $a_0 = \frac{\hbar^2 B}{12m\lambda^2}\xi$. The energy is then proportional to

$$\epsilon = -(\sqrt{1+\xi\eta}-1)^2\xi^2 + \eta^2/\xi. \qquad \text{(II.5.23)}$$

The contour plot of this function is presented in Figure II.5.3. After a straightforward calculation, one obtains $\xi_0 = 24$ and $\eta_0 = 1/3$ for the minimum in the energy. The corresponding parameter values of the original problem are $u_0 = \lambda/B$ and $a_0 = \hbar^2 B/m\lambda^2$.

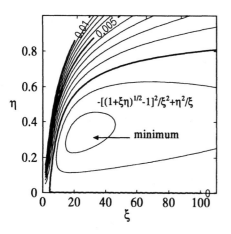

Fig. II.5.3. Contour plot of the energy function.

Apparently, in one dimension it is always better for the electron to induce a lattice distortion and thereby reduce its energy. The object thus created is called a *polaron* (the name refers to the lattice polarization induced by the electron). The size of the polaron, $a_0 = \hbar^2 B/m\lambda^2$, is larger for rigid lattices (large $B$) or weak electron–phonon interactions (small $\lambda$).

There is one important consideration remaining. According to our solution, the distortion of the lattice is $u_0/a_0 = m\lambda^3/\hbar^2 B^2$. However, if this number approaches unity (that is, the change of the interatomic distance is comparable to the lattice spacing), then the "harmonic approximation" for the elastic energy becomes invalid. In this case (referred to as a "small polaron" in the literature) it is more appropriate to take $u_0/a_0 = 1$ and minimize the energy using this condition. Similarly, it is meaningless to consider $a_0$ less than the lattice spacing of the material.

A more general, but still elementary discussion of the polaron problem can be found in Ref. [11] pp. 186–189.

## 5.4 Solution: Polaritons

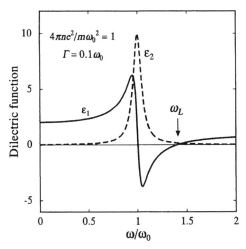

Fig. II.5.4. Real and imaginary part of the dielectric function of an insulator for representative values of $4\pi ne^2/m$ and $\Gamma$, as indicated in the figure. For this set of parameters, $\epsilon(\omega \to 0) = 2$ and $\epsilon(\omega \to \infty) = 1$. The longitudinal wave frequency is $\omega_L = 1.41\omega_0$.

The real and imaginary part of the dielectric function (Eq. I.8.18) is plotted in Figure 5.4. For $\Gamma \to 0$ the imaginary part of the dielectric function peaks sharply around $\omega = \omega_0$, and is zero otherwise. The solution of $\epsilon^* = 0$ (Eq. I.8.10) for longitudinal waves is

$$\omega = \omega_L = \sqrt{\omega_0^2 + \frac{4\pi ne^2}{m}} \,. \tag{II.5.24}$$

At this frequency the real part of $\epsilon^*$ is zero and the imaginary part is very small (or zero in the $\Gamma \to 0$ limit). According to the figure, $\epsilon_1$ also crosses zero at $\omega = \omega_0$. However, at this point $\epsilon_2$ is definitely nonzero.

There are two solutions for transverse electromagnetic waves. From Eq. I.8.9 we obtain

$$\omega^2 = \frac{c^2 k^2 + \omega_L^2}{2} \pm \frac{\sqrt{(c^2 k^2 + \omega_L^2)^2 - 4c^2 k^2 \omega_0^2}}{2} . \qquad (\text{II.5.25})$$

Here we have used $\omega_L$ as defined in Eq. II.5.24. For $ck \gg \omega_0$, one of the transverse wave solutions is $\omega_T = \omega_0$. (The other solution is $\omega = ck$, corresponding to propagating light waves.) The Lyddane–Sachs–Teller relation is proved by using $\omega_T = \omega_0$, Eq. II.5.24, and by taking the low and high frequency limits of Eqs. I.8.18.

For further discussions, see Ibach and Lüth [4] p. 251, Kittel [2] p. 304, or other textbooks. The polariton concept in cubic semiconductors is discussed by Cardona in Ref. [12] p. 441.

## 5.5 Hint: Excitons

Calculate the effective mass of the electrons and holes, then use the result in a "hydrogen atom" ground state energy formula. Compare the binding energy to $k_B T$.

Excitons are discussed, for example, by Kittel [2] p. 340, or Ibach and Lüth, [4] p. 269. Yu and Cardona [6] (pp. 266–282) provide a comprehensive introduction to the subject.

## 5.6 Solution: Holstein–Primakoff Transformation

We want to prove that the operators defined by Eq. I.5.23 satisfy the commutation relations given in Eq. I.5.14: $S^+ S^- - S^- S^+ = 2S^z$; $S^- S^z - S^z S^- = S^-$; and $S^z S^+ - S^+ S^z = S^+$. We will use the boson commutator $a^+ a - a a^+ = 1$, and we will take $\hbar = 1$. For the first equation we obtain:

$$\frac{1}{2S}[S^+, S^-] = \left[ \left( a^+ \sqrt{1 - \frac{a^+ a}{2S}} a \right), \left( \sqrt{1 - \frac{a^+ a}{2S}} a \right) \right]$$

$$= a^+ \left( 1 - \frac{a^+ a}{2S} \right) a - \sqrt{1 - \frac{a^+ a}{2S}} a a^+ \sqrt{1 - \frac{a^+ a}{2S}} \qquad (\text{II.5.26})$$

$$= \left( a^+ a - \frac{a^+ a^+ a a}{2S} \right) - \sqrt{1 - \frac{a^+ a}{2S}} (a^+ a - 1) \sqrt{1 - \frac{a^+ a}{2S}} .$$

For any function $F(a^+ a)$ the commutator $[F(a^+ a), a^+ a]$ is zero, and therefore the $a^+ a$ operator in the second term can be moved forward. Furthermore, we can replace $a^+ a^+ a a$ by $a^+ (1 + a a^+) a$ in the first term. This leads to

$$\frac{1}{2S}\left[S^+, S^-\right] = 1 - \frac{a^+ a}{S} = \frac{S^z}{2S} \,, \qquad (\text{II}.5.27)$$

as we wanted to prove.

The other two equations can be handled similarly. The operator identities $[AB, C] = A[B, C] + [A, C]B$ and $[A, BC] = [A, B]C + B[A, C]$ may be used to express the commutators.

## 5.7 Hint: Dyson–Maleev Representation

The solution is similar to Solution 5.6.

## 5.8 Solutions: Spin Waves

Let us write the Heisenberg model in terms of the spin-flip operators $S_i^+ = S_i^x + iS_i^y$ and $S_i^- = S_i^x - iS_i^y$ . We obtain

$$H = -J\sum_i S_i^z S_{i+1}^z + \frac{1}{2}(S_i^+ S_{i+1}^- + S_i^- S_{i+1}^+) \,. \qquad (\text{II}.5.28)$$

In the ground state of the ferromagnetic Heisenberg model, all spins are lined up in one direction (see, for example, Harrison [5] pp. 467–470). Let us choose this direction along the $z$ axis and denote the ground state as $|0\rangle$. It is easy to show that $H|0\rangle = -NS^2 J|0\rangle$. To construct a simple excitation, we flip a single spin at site $i$, $|i\rangle = S_i^-|0\rangle$. However, by using Eq. I.5.14, it is easy to show that this state is not a solution of the Schrödinger equation:

$$H|i\rangle = -NS^2 J|i\rangle + 2SJ|i\rangle - SJ(|i+1\rangle + |i-1\rangle) \,. \qquad (\text{II}.5.29)$$

Note that the result is valid for any $S$.

We are facing the familiar problem of finding the eigenvalues of the Hamiltonian with a strip of off-diagonal matrix elements, similar to the tight-binding electron problem or to the phonon problem. As always, the expansion in plane waves leads to a solution:

$$|k\rangle = \frac{1}{\sqrt{N}}\sum_j e^{ikR_j}|j\rangle \,. \qquad (\text{II}.5.30)$$

With this wavefunction the energies are given by

$$E(k) = E_0 + 2JS(1 - \cos ka) \,, \qquad (\text{II}.5.31)$$

where $E_0 = -NS^2 J$ is the ground-state energy.

For the Ising system we take the same approach, except the calculation becomes much simpler since we do not have to worry about the $S^+$ and

$S^-$ operators. In fact, the simple spin-flip state $S_i^-|0\rangle$ *is* a solution of the Schrödinger equation, with energy $E_0 + 2JS$, independent of the spin site flipped. Accordingly, a linear combination in the form of Eq. II.5.30 is also a solution, with an energy independent of $k$. The two solutions are plotted in Figure II.5.5.

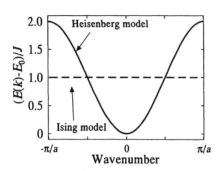

Fig. II.5.5. Magnon dispersion in the Heisenberg and Ising models.

Note that the magnon dispersion is parabolic at long wavelengths, similar to the electrons. With typical values for the exchange coupling $J$, the effective mass proves to be in the range of ten electron masses. However, we should keep in mind that magnons are bosons, whereas electrons are fermions, meaning their thermodynamic properties will be entirely different.

## 5.9 Solution: Spin Waves Again

After performing some operator algebra we obtain the result

$$H = \sum_{i,j} J_{ij} \left( S^2 - S(a_i^+ a_i + a_j^+ a_j - a_i^+ a_j + a_i a_j^+) \right) -$$
$$- \frac{1}{2} \left( a_i^+ a_i a_i a_j^+ + a_i^+ a_j^+ a_j a_j - 2a_i^+ a_i a_j^+ a_j \right) . \tag{II.5.32}$$

The quadratic terms are

$$H_0 = \sum_{i,j} J_{ij} \left( S^2 - S(a_i^+ a_i + a_j^+ a_j - a_i^+ a_j + a_i a_j^+) \right) . \tag{II.5.33}$$

This Hamiltonian is diagonalized by introducing the new operator $a_q$, defined in

$$a_j = \frac{1}{\sqrt{N}} \sum_q e^{iqR_i} a_q , \tag{II.5.34}$$

and a similar equation for $a^+$. At this point it is practical to restrict the calculation to nearest-neighbor interactions. We obtain

$$H_0 = -JS \sum_q \sum_\ell (\cos q\ell - 1) a_q^+ a_q \, , \qquad \text{(II.5.35)}$$

where the $\ell$ vector points to nearest-neighbors.

This result is meaningful for $J > 0$ (ferromagnetic coupling) and at low temperatures. For $J > 0$, the ground state $|0\rangle$ has all spins pointing in one direction. Let us choose this direction as the $z$ direction for the spin operator. It is easy to show that $H_0|0\rangle = 0$. The operator $a_q^+$ creates an elementary excitation (magnon) with an energy $E(q) = -JS \sum_q \sum_\ell (\cos q \cdot \ell - 1)$. For a linear chain this result is identical to Solution 5.8. As long as there is only one magnon in the system, the fourth-order terms in the original Hamiltonian, Eq. II.5.32, are all exactly zero. At low temperatures, when only a few magnons are excited, the solution is still acceptable. The fourth-order terms can be viewed as interactions between the magnons, and they become important at higher temperatures.

Note that similar results are obtained if the Holstein–Primakoff transformation (see Problem 5.6) is used to express the spin operators in terms of bosons. Mattis [8] pp. 149–156 discusses the solution of the ferromagnetic Heisenberg model in detail.

## 5.10 Solution: Anisotropic Heisenberg Model

Note that for $J = 0$ and spin $1/2$ the anisotropic Heisenberg model reduces to the Ising model; taking $J' = 0$ leads to the so-called "$x - y$ model."

The solution is similar to the solution of Problem 5.8, except we first rewrite the Hamiltonian in the form of

$$H = -J \sum_i S_i S_{i+1} + \frac{(J' - J)}{2} \sum_i S_i^z S_{i+1}^z \, . \qquad \text{(II.5.36)}$$

The spin-wave wavefunctions are given by

$$|k\rangle = \frac{1}{\sqrt{N}} \sum_j e^{ikR_j} |j\rangle \, , \qquad \text{(II.5.37)}$$

where the state $|j\rangle$ is created from the ferromagnetically ordered state $|0\rangle$ by reducing the $z$ component of the spin at site $j$, and $R_j = aj$ is the site position. The magnon energy will be

$$E(k) = E_0 + 2JS(1 - \cos ka) + 2S(J' - J) \, . \qquad \text{(II.5.38)}$$

(a) For $J' > J$ the result suggests that the lowest magnon energy is $2S(J' - J)$.

(b) For $J' < J$ we find the surprising result that the system with one spin flip may have *lower* energy than the ordered state. Therefore the ordered state may not be the ground state. Finding the ground state in this case is a rather nontrivial task. A detailed discussion, including the exact solution for the one-dimensional $x - y$ model is discussed by Mattis [8] pp. 159–163.

## 5.11 Hint: Solitons

(a) The solution is tested by calculating the second derivatives and substituting them into Eq. I.5.27. The parameter $\gamma$ turns out to be $\gamma = 1/\sqrt{1 - v^2}$. The function described by Eq. I.5.28 is plotted in Figure II.5.6. The width is $\Delta \approx \pi/m$; the energy is $E_0 = 8m\gamma$. The resemblance of $\gamma$ to the factor appearing in the theory of special relativity is no coincidence: Eq. I.5.27 is invariant under the Lorentz transformation.

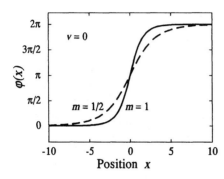

**Fig. II.5.6.** Single-soliton solution of Eq. I.5.27, as described by Eq. I.5.28.

(b) The proof of the solution is similar to that of part (a). Here $\gamma = 1/\sqrt{1 + a^2}$ and $E = 16m\gamma$. Since $\gamma < 1$, the energy is always less than $2E_0 = 16m$.

The are two other "two-soliton" solutions:

$$\varphi = 4\tan^{-1} \frac{1}{u} \frac{\sinh(m\gamma ut)}{\cosh(m\gamma x)} \qquad (\text{II.5.39})$$

and

$$\varphi = 4\tan^{-1} \frac{1}{u} \frac{\sinh(m\gamma x)}{\cosh(m\gamma ut)}, \qquad (\text{II.5.40})$$

describing free soliton–antisoliton and soliton–soliton pairs, respectively.

# 6 Thermodynamics of Noninteracting Quasiparticles

## 6.1 Solution: Specific Heat of Metals and Insulators

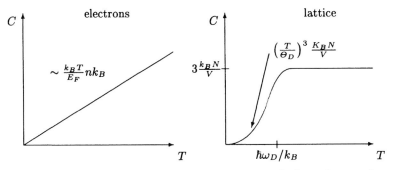

**Fig. II.6.1.** Temperature dependence of the specific heat due to electrons and lattice vibrations.

The electronic and lattice contributions to the specific heat are shown in Figure II.6.1. For electrons,

$$c_e \sim \frac{k_B T}{E_F} n k_B \ . \tag{II.6.1}$$

For an approximately isotropic three-dimensional insulating crystal the specific heat varies with temperature as sketched in the right hand panel of Figure II.6.1. Above the temperature corresponding to the Debye frequency the specific heat becomes temperature independent. At low $T$,

$$c_{ph} \sim \left(\frac{T}{\Theta_D}\right)^3 \frac{N k_B}{V} \ . \tag{II.6.2}$$

Typical values are $E_F/k_B = T_F \sim 10{,}000$ K, $\Theta_D = \hbar\omega_D/k_B \sim 100$ K. Since $n \sim N/V$, at $T = 1$ K we get

$$\frac{c_{ph}}{c_e} = T^2 \frac{T_F}{\Theta_{rmD}^2} = 0.01 \tag{II.6.3}$$

187

At $T = 500$ K we obtain a ratio of

$$\frac{c_{ph}}{c_e} = \left(\frac{500}{10,000}\right)^{-1} \sim 20 \; . \tag{II.6.4}$$

## 6.2 Solution: Number of Phonons

We will use Eq. I.6.25 with the density of states from Eq. I.4.7, and the Bose–Einstein distribution function, Eq. I.6.22. The chemical potential is $\mu = 0$. For each branch $s$ the integration is over a finite range of energies, from $\mathcal{E} = 0$ to $\mathcal{E} = \hbar c_s k_D$. We obtain the number of phonons in each mode $s$ to be

$$N_s = 3N_a \left(\frac{k_B T}{\hbar c_s k_D}\right)^3 \int_0^{x_0} \frac{x^2}{e^x - 1} \, dx \; , \tag{II.6.5}$$

where $k_D$ is the Debye wavenumber, $N_a$ is the number of atoms in the solid and the upper limit of the integration is $x_0 = \hbar c_s k_D / k_B T$. The total number of phonons is $N = N_1 + N_2 + N_3$.

For low temperatures, the upper limit of the integral diverges, and the integral converges to a constant, independent of $s$. Therefore the number of phonons varies as

$$N \propto \left(\frac{1}{c_1^3} + \frac{1}{c_2^3} + \frac{1}{c_3^3}\right) \left(\frac{k_B}{\hbar k_D}\right)^3 T^3 \; . \tag{II.6.6}$$

For high temperatures, $x$ remains small within the range of integration and the distribution function can be expanded into a Taylor series. The integral can be performed, yielding

$$N \propto \left(\frac{1}{c_1} + \frac{1}{c_2} + \frac{1}{c_3}\right) \frac{k_B}{\hbar k_D} T \; . \tag{II.6.7}$$

Note that the coefficient in front of the temperature dependent term contains an "averaged" sound velocity, but the average is different at low and at high temperatures. If we take the three sound velocities to be approximately equal, the expressions are further simplified to $N \propto (T/\theta_D)^3$ at low $T$, and $N \propto (T/\theta_D)$ at high $T$. Here $\theta_D = \hbar c k_D / k_B$ is the Debye temperature. For this particular case the results of the numerical integration of Equation II.6.5 is shown in Figure II.6.2.

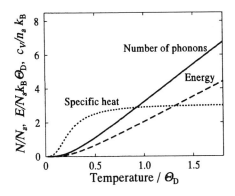

Fig. II.6.2. Temperature dependence of the number of phonons,
the energy of the phonons, and the
phonon contribution to the specific
heat for a simple solid with one
atom per unit cell in the Debye approximation. $N_a$ is the number of
atoms in the solid; all three sound
velocities were taken to be equal.

# 6.3 Solution: Energy of the Phonon Gas

The energy is calculated similar to the calculation of the particle number in
Solution 6.2. The energy is expressed by Eq. I.6.24. The Debye density of
states, Eq. I.4.7, and the Bose–Einstein distribution function, Eq. I.6.22 are
inserted. To further simplify the calculation, we assume that all three sound
velocities are equal to $c$. We obtain

$$E = 9N_a k_B T \left(\frac{T}{\theta_D}\right)^3 \int_0^{x_0} \frac{x^3}{e^x - 1}\, dx \,, \qquad (\text{II}.6.8)$$

where $\theta_D$ is the Debye temperature, $N_a$ is the number of atoms in the solid
and the upper limit of integration is $x_0 = \theta_D/k_B T$. A factor of 3 was included
for the three phonon modes. The result of a numerical integration is shown
in Figure II.6.2.

For low temperatures, the upper limit of the integral diverges, and the
integral becomes a constant. Therefore the energy will increase like $E \propto
T^4$. In this regime the phonon system is similar to photons, and the $\sim T^4$
temperature dependence is similar to the blackbody radiation formula. For
high temperatures, $x$ remains small within the range of integration and the
distribution function can be expanded into a Taylor series. The integral can
be performed, yielding $E = N_a k_B T$.

The specific heat is obtained by taking the temperature derivative of the
energy density, $E/V$, where $E$ is given in Eq. II.6.8. The result is[1]

$$c_V = 9n_a k_B \left(\frac{T}{\theta_D}\right)^3 \int_0^{x_0} \frac{x^4 e^x}{(e^x - 1)^2}\, dx \,, \qquad (\text{II}.6.9)$$

---

[1] See also Ashcroft and Mermin [1] p. 457, or Kittel [2] p. 137.

also shown in Figure II.6.2. Here $n_a = N_a/V$ is the density of the atoms. At low temperatures the specific heat behaves as $c_V = n_a k_B (12\pi^4/5)(T/\theta_D)^3$. At high temperatures the specific heat is $c_V = 3n_a k_B$.

The high-temperature behavior of phonons satisfies the *Dulong–Petit law*: each atom has six degrees of freedom and each degree of freedom carries $\frac{1}{2}k_B T$ energy.

## 6.4 Solution: Bulk Modulus of Phonon Gas

Let us first express the volume dependence of the energy of the phonon gas. The energy is

$$E = \sum_k f(\hbar\omega_k)\hbar\omega(k) , \qquad (\text{II}.6.10)$$

where $f$ is the Bose function, and $k$ is the phonon wavenumber. Therefore

$$\frac{\partial E}{\partial V} = \sum_k f\hbar\frac{d\omega}{dV} + \sum_k \frac{df}{d\omega}\frac{d\omega}{dV}\hbar\omega$$

$$= -\frac{\gamma}{V}\sum_k \hbar\omega\left[f + \frac{df}{d\omega}\omega\right] . \qquad (\text{II}.6.11)$$

Note that the Grüneisen relation assumes that each phonon mode shifts by exactly the same relative amount as the volume is changed. This approximation allowed us to obtain a relatively simple result here.

We proceed to the calculation of the thermodynamic properties. According to the definitions of the pressure and bulk modulus (Eqs. I.6.5 and I.6.10), we need the volume derivative of the free energy, $\partial F/\partial V$. The free energy is given in Eq. I.6.27, and therefore

$$\frac{\partial F}{\partial V} = \frac{\partial}{\partial V}\{-T\int\frac{1}{T'^2}E\,dT'\} = -T\int\frac{1}{T'^2}\frac{\partial E}{\partial V}\,dT' . \qquad (\text{II}.6.12)$$

When we substitute $dE/dV$ from Eq. II.6.11, we can exploit a general property of the Bose function:

$$\frac{df}{d\omega} = -\frac{T}{\omega}\frac{df}{dT} \qquad (\text{II}.6.13)$$

The volume derivative of the free energy is

$$\frac{\partial F}{\partial V} = \frac{\gamma}{V}T\int\frac{1}{T'^2}\sum_k \omega\left[f - T'\frac{df}{dT'}\right]dT' . \qquad (\text{II}.6.14)$$

If we interchange the $\int$ and the $\sum_k$, then we can perform an integration by parts. We find that the $\int(1/T'^2)f\omega\,dT'$ term is canceled. The remaining term is identical to the Eq. II.6.10, the energy. The final result is

$$p = -\frac{\partial F}{\partial V} = \gamma \frac{E}{V} \ , \tag{II.6.15}$$

and the bulk modulus is

$$B = -V \frac{\partial p}{\partial V} = (\gamma^2 + \gamma) \frac{E}{V} \tag{II.6.16}$$

In the Debye approximation, we obtain at low temperatures (see Problem 6.3) $E \approx 9 N_a k_B T (T/\theta_D)^3 (\pi^4/15)$, where $N_a$ is the number of atoms, and $\theta_D$ is the Debye temperature. At high temperatures the energy is $E \approx 3 N_a k_B T$.

Typical values of $B$ for solids are in the range of $10^{11} - 10^{12}$ dyn/cm$^2$, approximately independent of the temperature. Let us estimate the value of bulk modulus from Eq. II.6.16 at 500 °C. Assuming that the Debye temperature is less than this temperature, we get $B \approx 3(\gamma^2 + \gamma) n_a k_B T = 3(\gamma^2 + \gamma) \ 10^{23}$ cm$^{-3}$ $6 \times 10^{-14}$ erg $\approx 4 \times 10^9$ dyn/cm$^2$. Here we used a typical density of $n_a = 10^{23}$ cm$^{-3}$, and a $\gamma$ in the order of 1.

We can conclude that at low temperatures the phonons do not contribute to the bulk modulus, but near to the melting point the phonon contribution may become significant.

## 6.5 Hint: Phonons in One Dimension

See Solutions 6.2 and 6.3.

## 6.6 Solution: Electron–Hole Symmetry

Using electron–hole symmetry, the upper and lower energy bands of the semiconductor will be symmetric around the middle of the gap [this is also evident from the upper band being defined as $E_0 - E(k)$]. The density of states will therefore look similar to what is sketched in Figure II.6.3. The total density of states is given by $g(E) = g_1(E) + g_1(E_0 - E)$.

(a) To calculate the chemical potential, we begin with the condition

$$N = V \int g(E) f(E) \ \mathrm{d}E = \int [g_1(E) + g_1(E_0 - E)] f(E) \ \mathrm{d}E \ . \tag{II.6.17}$$

Note that the Fermi function $f(E)$ has the property

$$f(E - \mu) = 1 - f(\mu - E) \ . \tag{II.6.18}$$

And note that at $T = 0$, we obtain

$$N = V \int g_1(E) \ \mathrm{d}E \ . \tag{II.6.19}$$

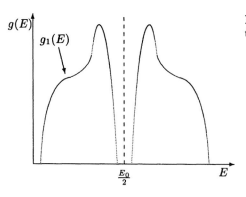

**Fig. II.6.3.** Density of states for a two-band semiconductor.

Consequently,

$$\int g_1(E) \, dE = \int [g_1(E) + g_1(E_0 - E)] \, f(E - \mu) \, dE ,\tag{II.6.20}$$

and

$$\int g_1(E) \, dE = \int [g_1(E) + g_1(E_0 - E)] \, (1 - f(\mu - E) \, dE ,\tag{II.6.21}$$

which can be rewritten as

$$-\int g_1(E) \, dE = -\int [g_1(E) + g_1(E_0 - E)] \, f(\mu - E) \, dE .\tag{II.6.22}$$

By putting Eqs. II.6.20 and II.6.22 together, we find the condition

$$\int g_1(E) f(E - \mu) \, dE = \int g_1(E) f(\mu - E) \, dE .\tag{II.6.23}$$

This equation can only be satisfied if $\mu = E_0/2$, independent of the temperature. Therefore the chemical potential is $E_0/2$ at *all* temperatures!

(b) For low temperatures we can expand the Fermi function, $f(E)$, because the density of states, $g(E)$, is zero over an energy range of $2\Delta$ centered around $E_0/2$ (see Figure II.6.3). Defining $\Delta E = E - E_0/2 \gg k_B T$, we obtain

$$f(E) = \begin{cases} 1 - e^{\frac{-|\Delta E|}{k_B T}} & \text{for} \quad \Delta E < 0 \\ e^{\frac{-\Delta E}{k_B T}} & \text{for} \quad \Delta E > 0 \end{cases} .\tag{II.6.24}$$

## 6.7 Solution: Entropy of the Noninteracting Electron Gas

We start with the result presented in Eq. I.6.26 (except, to simplify the calculation, we keep the chemical potential as a variable):

$$S(T, V, \mu) = \int_{T_0}^{T} \frac{1}{T'} \frac{\partial E(T', V, \mu)}{\partial T'} \, dT' \, . \qquad (\text{II.6.25})$$

Inserting the energy functions from Eq. I.6.24, Eq. II.6.25 becomes

$$S(T, V, N) = \int g(E)(E - \mu) \left( \int_{T_0}^{T} \frac{1}{T'} \frac{\partial f(T', \mu)}{\partial T'} \, dT' \right) dE \, . \qquad (\text{II.6.26})$$

The question is: What should we do with the integral in the parentheses

$$I = \int \frac{1}{T'} \frac{df(T')}{dT'} \, dT' \, ? \qquad (\text{II.6.27})$$

Since the only relevant variable is the temperature, we replaced the partial derivative symbols with full derivatives and we will worry about the limits of the integration later. Note that the integral becomes simple if we invert $f(T)$ to obtain $T = T(f)$ and introduce $f$ as a new variable,

$$I = \int \frac{1}{T(f)} \, df \, , \qquad (\text{II.6.28})$$

where, from the definition of the Fermi function given in Eq. I.6.23, we have $1/T = \frac{k_B}{E - \mu} \log \frac{1-f}{f}$. Therefore

$$
\begin{aligned}
I &= \frac{k_B}{E - \mu} \int \log \frac{1 - f}{f} \, df \\
&= \frac{k_B}{E - \mu} [f \log f + (1 - f) \log(1 - f)] \, . \qquad (\text{II.6.29})
\end{aligned}
$$

In the last step we used standard calculus methods to determine the primitive function. Substituting this result into Eq. II.6.26, we obtain

$$
\begin{aligned}
S(T, V, \mu) &= \int g(E)(E - \mu) \frac{k_B}{E - \mu} f \log f + (1 - f) \log(1 - f) \, dE \\
&= k_B \int g(\mathcal{E}) \left( f \log f + (1 - f) \log(1 - f) \right) dE \, . \qquad (\text{II.6.30})
\end{aligned}
$$

This answer can be viewed as the weighted average of the configuration entropies of the electrons and holes. The volume dependence is carried within $g(E) = g(E, V)$; the temperature and the chemical potential appear in the Fermi function $f(T, \mu)$. According to the third law of thermodynamics, the entropy must be zero at $T = 0$. This condition is satisfied by the above formula. At zero temperature, within the first term either $f$ or $\log f$ is zero and within the second term either $(1 - f)$ or $\log(1 - f)$ is zero, therefore the integrand is zero.

## 6.8 Solution: Free Energy with Gap at the Fermi Energy

(a) At zero temperature the energy of the electron gas is calculated from Eq. I.6.24:

$$E = V \int_0^{\mathcal{E}_F} g(\mathcal{E})\mathcal{E}d\mathcal{E} \ , \tag{II.6.31}$$

where we have assumed that the lower edge of the band is at $\mathcal{E} = 0$. Similarly, the total number of electrons is given by

$$N = V \int_0^{\mathcal{E}_F} g(\mathcal{E})d\mathcal{E} \approx g_0\mathcal{E}_F; . \tag{II.6.32}$$

A simple calculation proves that the total number of states in the lower band does not depend upon the value of the gap, and therefore the Fermi energy will not change as the gap opens. Let us chose the zero energy to be at the Fermi energy; $\mathcal{E}_F = 0$. Using this energy scale the band extends from $-\mathcal{E}_F = 0$ to $W - \mathcal{E}_F$. The change in the total energy becomes

$$\begin{aligned} \delta E &= V \int_{-\mathcal{E}_F}^{-\Delta} g_0 \frac{-\mathcal{E}^2}{\sqrt{\mathcal{E}^2 - \Delta^2}} d\mathcal{E} - V \int_{-\mathcal{E}_F}^0 g_0 \mathcal{E} d\mathcal{E} \\ &= V \int_{-W/2}^{-\Delta} g_0 \left( \frac{\mathcal{E}}{\sqrt{-\mathcal{E}^2 - \Delta^2}} - 1 \right) \mathcal{E} \, d\mathcal{E} - g_0 V \Delta; . \end{aligned} \tag{II.6.33}$$

The limits of the integration extend over the energy range where the density of states is nonzero, and within this range $-\mathcal{E}$ is positive. Note that for energies much below the Fermi energy, the integrand rapidly converges to zero. Therefore, as long as the Fermi energy is far from the band edge, the choice of the lower limit will not influence the result in any important way.

The integration can be performed exactly. For $W \gg \Delta$ the leading term is

$$\delta E \approx \frac{1}{2} g_0 V \Delta^2 \log \Delta/\mathcal{E}_F \approx \frac{1}{2} \frac{\Delta^2}{W} \log \Delta/W \ , \tag{II.6.34}$$

since $g_0 \approx 1/W$ and $W$ and $\mathcal{E}_F$ are of the same order of magnitude. A similar result was obtained in Solution 3.8.

Due to the logarithmic term in this expression, the electron system gains a significant amount of energy as the gap opens. As a consequence of this energy gain, the metallic state with a zero gap at the Fermi level may become unstable, and a *Peierls transition* may happen. This issue will be discussed in Problem 9.6. Note that the logarithmic term is due to the singular density of states, which, in turn, is due to the one-dimensional character of the electron gas (see Problem 4.5). It is all the more interesting that in real (quasi-one-dimensional) materials, like $TaS_3$, $NbSe_3$, or $K_{0.3}MoO_3$ the divergence of the density of states is strong enough to produce some of the effects described here.

(b) The entropy is calculated in Problem 6.7. From Eq. II.6.30, the entropy per unit length is:

$$S/L = k_B \int g(\mathcal{E}) \left( f \log f + (1 - f) \log(1 - f) \right) d\mathcal{E} . \qquad (\text{II.6.35})$$

Note that $f(\mathcal{E}) = 1 - f(-\mathcal{E})$, and the density of state has electron–hole symmetry, $g(\mathcal{E}) = -g(\mathcal{E})$. Therefore the contribution of the two terms in the integrand will be the same.

$$S = 2Lk_B \int g(\mathcal{E}) f \log f \, d\mathcal{E} . \qquad (\text{II.6.36})$$

The full expression for the free energy will be

$$F = E - TS = L \int g f (\mathcal{E} - 2k_B T \log f) d\mathcal{E} . \qquad (\text{II.6.37})$$

It is easy to show that $\log f \approx -\mathcal{E}/k_B T$ for $\mathcal{E} \gg k_B T$ and $\log f \approx 0$ for $\mathcal{E} \ll k_B T$. Due to the density of states and the Fermi function in the integrand, the limits of integrations will run from $-\mathcal{E}_F$ to $-\Delta$. In this energy range the entropy term is always negligible.

## 6.9 Solution: Bulk Modulus at $T = 0$

The zero temperature bulk modulus of the free electron gas is calculated in the solution to Problem 6.10, and it is discussed by several textbooks, including Ashcroft and Mermin [1] p. 39, and Kittel [2] p. 179. The result is

$$B = \frac{10}{9} \frac{E}{V} = \frac{\hbar^2}{3m} (3\pi^2)^{2/3} \left( \frac{N}{V} \right)^{5/3} \qquad (\text{II.6.38})$$

In contrast to the phonon contribution (see Problem 6.4), the bulk modulus due to the electrons does not disappear at zero temperature.

## 6.10 Solution: Temperature Dependence of the Bulk Modulus

According to Eq. I.6.10 the bulk modulus is related to the compressibility, $\kappa$, and is defined as

$$B = \frac{1}{\kappa} = -V \frac{\partial p}{\partial V}, \qquad (\text{II.6.39})$$

where the pressure $p$ is related to the free energy as

$$p = -\frac{\partial F}{\partial V}. \qquad (\text{II.6.40})$$

To obtain the free energy of the electron gas, we use Eq. I.6.27:

$$F(T, V, N) = -T \int_0^T \frac{1}{T'^2} E(T', V, N) \, dT' + \frac{T}{T_0} E(T_0) \qquad \text{(II.6.41)}$$

The energy, $E$, can be calculated using the Bethe–Sommerfeld expansion, Eq. I.6.30, yielding

$$E = E_0(V) + \frac{\pi^2}{6} V (k_{\mathrm{B}} T)^2 g(\mathcal{E}_{\mathrm{F}}) , \qquad \text{(II.6.42)}$$

where $E_0(V)$ is the ($T = 0$) ground-state energy. For a three-dimensional free-electron gas of energy $E = \hbar^2 k^2 / 2m$, Eq. I.4.4 leads to the density of states[2]

$$g(E_{\mathrm{F}}) = g(k_{\mathrm{F}}) = \frac{m k_F}{\hbar^2 \pi^2} , \qquad \text{(II.6.43)}$$

and the Fermi wavenumber, $k_F$, is defined by the condition

$$N = 2 \int \frac{d^3 k}{(2\pi/L)^3} \qquad \Longrightarrow \qquad k_F = \left( 3\pi^2 \frac{N}{V} \right)^{1/3} . \qquad \text{(II.6.44)}$$

Note that in Eq. II.6.42 the energy is expressed as a function of $E = E(T, V, N)$; the temperature and volume dependence appear explicitly, while the volume $V$ and electron number $N$ enter into the $g(E_{\mathrm{F}})$ factor. This is exactly what we need to evaluate the integral in Eq. II.6.41. In the $T_0 \to 0$ limit we obtain

$$F = E_0 - \frac{\pi^2}{6} V (k_{\mathrm{B}} T)^2 g . \qquad \text{(II.6.45)}$$

The temperature-dependent correction to the free energy is negative, whereas the energy (Eq. II.6.42) has a positive term.

To calculate the bulk modulus, $B$, we have to make the volume dependence of $F$ explicit. Inserting $g = g(V, N)$, we get

$$E_0(V) = E_{\mathrm{tot}} = 2 \int \frac{d^3 k}{(2\pi/L)^3} \frac{\hbar^2 k^2}{2m} = \frac{\hbar^2 k_F^5}{10 m \pi^2} V \qquad \text{(II.6.46)}$$

where the 2 in front of the integral is for spin, $V (= L^3)$ is the volume. Finally

$$F = \frac{\hbar^2}{10 m \pi^2} (3\pi^2 N)^{5/3} V^{-2/3} - \frac{\pi^2}{6} (k_{\mathrm{B}} T)^2 \frac{m}{\hbar^2 \pi^2} (3\pi^2 N)^{1/3} V^{2/3}. \qquad \text{(II.6.47)}$$

We can then calculate the pressure using Eq. II.6.40,

$$p = \frac{2}{3} \frac{\hbar^2}{10 m \pi^2} (3\pi^2 N)^{5/3} V^{-5/3} + \frac{2}{3} (k_{\mathrm{B}} T)^2 \frac{m}{6\hbar^2} (3\pi^2 N)^{1/3} V^{-1/3} . \qquad \text{(II.6.48)}$$

---

[2] Also see Ashcroft and Mermin [1] pp. 34–36, or Ibach and Lüth [4] pp. 86–88.

Note that the pressure of the free-electron gas increases with temperature, although the temperature dependence is far from the ideal gas law ($pV = Nk_BT$). The bulk modulus is

$$B = \frac{10}{9}\frac{E}{V} = \frac{\hbar^2}{3m}(3\pi^2)^{2/3}\left(\frac{N}{V}\right)^{5/3} + \frac{2}{9}(k_BT)^2\frac{m}{6\hbar^2}(3\pi^2N)^{1/3}V^{-1/3}.$$

(II.6.49)

Only the second term is temperature-dependent, so the leading temperature-dependent term is

$$\frac{2}{54}\frac{m}{\hbar^2}(k_BT)^2\left(3\pi^2\frac{N}{V}\right)^{1/3}.$$

(II.6.50)

Since $N/V = n$, this term is proportional to $n^{1/3}$, and therefore it is inversely proportional to the average distance between the particles.

## 6.11 Solution: Chemical Potential of the Free-Electron Gas

The Bethe–Sommerfeld expansion, Eq. I.6.31, works well in one and three dimensions. We need to $g'(E_F)$ and $g(E_F)$ in order to calculate the chemical potential, $\mu$.

$$1 - D \qquad\qquad\qquad 3 - D$$

$$g(E) = \frac{m}{\pi\hbar^2}\sqrt{\frac{\hbar^2}{2m}}\frac{1}{\sqrt{E}} \qquad\qquad g(E) = \frac{m}{\pi^2\hbar^2}\sqrt{\frac{2m}{\hbar^2}}\sqrt{E}$$

$$= A_1\frac{1}{\sqrt{E}} \qquad\qquad\qquad = A_3\sqrt{E}$$

$$g'(E) = \frac{\delta g}{\delta E} = -\frac{1}{2}A_1E^{-3/2} \qquad g'(E) = \frac{\delta g}{\delta E} = \frac{1}{2}A_3E^{-1/2}$$

$$g'(E)/g(E) = -\frac{1}{2}E^{-1} \qquad\qquad g'(E)/g(E) = \frac{1}{2}E^{-1}$$

$$\rightarrow\quad \mu = E_F + \frac{\pi^2}{12}\frac{(k_BT)^2}{E_F} \qquad \rightarrow\quad \mu = E_F - \frac{\pi^2}{12}\frac{(k_BT)^2}{E_F}$$

In two dimensions, $g'$ is zero. Therefore we either try to include higher-order terms in the Bethe–Sommerfeld expansion, or go back to the original definition of the chemical potential. We'll choose the latter route; we will find out that including higher-order terms would not work anyway. First,

$$N = 2\int\frac{d^2k}{(2\pi/L)^2}f(\mathcal{E}) .$$

(II.6.51)

Then by using

$$\mathcal{E} = \frac{\hbar^2 k^2}{2m}, \qquad d^2 k = 2\pi k \; dk, \qquad d\mathcal{E} = \frac{\hbar^2}{2m} 2k \; dk \qquad \text{(II.6.52)}$$

we obtain

$$N = \frac{Vm}{\pi\hbar^2} \int_0^\infty \frac{d\mathcal{E}}{e^{(\mathcal{E}-\mu)/k_B T} + 1} . \qquad \text{(II.6.53)}$$

This integral is easily solved (or looked up), resulting in

$$N = \frac{Vm}{\pi\hbar^2} T \ln\left(1 + e^{\mu/T}\right) . \qquad \text{(II.6.54)}$$

At $T = 0$, $\ln\left(1 + e^{\mu/T}\right) = \mu/T$, and $\mu = \mathcal{E}_F$. Therefore

$$N = \frac{Vm\mu}{\pi\hbar^2} \qquad \Longrightarrow \qquad \mathcal{E}_F = \frac{N}{V}\frac{\pi\hbar^2}{m}. \qquad \text{(II.6.55)}$$

We can express $\mu(T)$ as

$$\mu = \mathcal{E}_F + k_B T \ln\left(1 - e^{\mathcal{E}_F/k_B T}\right) . \qquad \text{(II.6.56)}$$

One can see that this equation behaves nonanalytically around $T = 0$, which is why the Bethe–Sommerfeld expansion would not work in two dimensions.

## 6.12 Solution: EuO Specific Heat

In metals the low-temperature specific heat has a linear temperature dependence. Since the exponent in the temperature dependence of the experimentally determined specific heat is larger than 1, at low enough temperatures the linear term would dominate. EuO must be an insulator.

   The low-temperature excitations may not be fermions, since that would lead to a linear term in the specific heat (as long as the temperature is much lower than the Fermi temperature).

   The energy of the system is

$$U = \int_0^{E_0} g(E)Ef(E) \; dE, \qquad \text{(II.6.57)}$$

where $f(E) = 1/(e^{E/k_B T} - 1)$ is the Bose factor. The lower limit of the integration is determined by the density of states: $g(E) = 0$ for $E < 0$. Strictly speaking, the upper limit is also determined by the density of states, since there are no states above a certain energy, $E^*$. However, at low temperatures (i.e., $k_B T \ll E^*$) the Bose factor imposes a strong, exponential cutoff at energies well below $E^*$. Accordingly, at low enough temperatures we can replace the upper limit with an arbitrary energy $E_0$, such that $k_B T \ll E_0 \ll$

$E^*$. Therefore the thermal properties of the system are determined by the low energy part of $g(E)$.

Let us search for a spherically symmetric spectrum in the form of $E = k^\alpha$, where $k = |\, k\, |$ and $\alpha > 0$ is an exponent to be determined. In three dimensions the density of states will be $g(E) \sim \int \frac{dS}{|dE/dk|} \sim k^2 k^{1-\alpha} \sim k^{3-\alpha} \sim E^{3/\alpha-1}$.

The specific heat can be calculated as follows:

$$c_V = \mathrm{d}u/\mathrm{d}T \;=\; \int_0^{E_0} g(E) E \frac{\mathrm{d}}{\mathrm{d}T} f(E)\, \mathrm{d}E$$

$$\sim \int_0^{E_0} E^{3/\alpha-1} E^2/(kB * T^2) f'\, \mathrm{d}E\,, \qquad (\mathrm{II}.6.58)$$

where $f' = f'(E/k_B T)$ is the derivative of the Bose function. The integration is from zero to the upper energy cutoff of the spectrum. By introducing the new variable $x = E/k_B T$, one obtains

$$c_V = T^{3/\alpha} \int_0^{E_0/k_B T} f'(x) x^{3/\alpha+1}\, \mathrm{d}x\,. \qquad (\mathrm{II}.6.59)$$

As the temperature goes to zero, the upper limit of the integral diverges, and the integral becomes independent of the temperature. Therefore the experimentally observed specific heat exponent corresponds to $\alpha = 2$; the energy dispersion is quadratic in $k$.

Magnons (see Problem 5.8) are bosons with an $E \sim k^2$ dispersion. Magnons are the source of the specific heat measured on EuO.

## 6.13 Hint: Magnetization at Low Temperatures

First, for low temperatures, only the low energy magnons are excited and the magnon dispersion[3] can be expanded as $E(q) = 2JS(3 - \cos q_x a - \cos q_y a - \cos q_z a) = JSa^2 q^2$. In three dimensions this quadratic dispersion leads to a density of states $g(E) \propto \sqrt{E}$ (see Table I.4.1 and Problem 4.1). The number of magnons is determined by Eq. I.6.25. At low temperatures the Bose function will cut off rapidly. We can use the low energy expansion of the density of states and extend the integration to $E \to \infty$. Introducing the new variable $x = E/k_B T$ will enable us to determine the temperature dependence of the number of magnons as $N_m \propto (k_B T/JS)^{3/2}$.

Each magnon corresponds to one spin-flip (see Solution 5.8). The magnetization is therefore $M(T) = NS - N_m$, where $N$ is the total number of spins. Substituting the result derived earlier, the magnetization will decrease as $\Delta M(T) \propto T^{3/2}$. This is "Bloch's $T^{3/2}$" law.[4]

---

[3] the magnon dispersion was determined in Solution 5.9.
[4] See, for example, Ashcroft and Mermin [1] p. 708, Kittel [2] p. 471, or Ibach and Lüth [4] p. 151.

## 6.14 Solution: Electronic Specific Heat

According to the result obtained from the Bethe–Sommerfeld expansion, the specific heat of the electrons depends on the density of states at the Fermi level (see Eq. I.6.32):

$$c_V = \frac{\pi^2}{3} k_B^2 T g(E_F) \qquad (\text{II.6.60})$$

The density of states is given by Eq. I.4.4 as

$$g = 2 \int \frac{dS}{8\pi^3} \frac{1}{|(\partial E(k)/\partial k)|}, \qquad (\text{II.6.61})$$

where the factor of two stands for the spin degeneracy. We also need the density of electrons, related to the volume occupied the Fermi surface:

$$n = 2 \int \frac{d^3k}{8\pi^3} = \int_0^{E_F} \frac{dE}{4\pi^3} g(E) . \qquad (\text{II.6.62})$$

For free-electrons, $|\nabla E| = \hbar^2 k/m$ and the density of states is

$$g = \int \frac{dS}{4\pi^3} \frac{m}{\hbar^2 k} = \frac{\sqrt{2}}{\pi} \left(\frac{m}{\hbar^2}\right)^{3/2} \sqrt{E} , \qquad (\text{II.6.63})$$

In order to calculate $g$ in the tight-binding model, we need the derivatives of $E(k)$:

$$\frac{\partial E}{\partial k_x} = E_1 a \sin k_x a \qquad (\text{II.6.64})$$

$$\frac{\partial E}{\partial k_y} = E_2 b \sin k_y b \qquad (\text{II.6.65})$$

$$\frac{\partial E}{\partial k_z} = E_3 c \sin k_z c \qquad (\text{II.6.66})$$

$$|\nabla E| = \sqrt{(E_1 a)^2 \sin^2 k_x a + (E_2 b)^2 \sin^2 k_y b + (E_3 c)^2 \sin^2 k_z c} \qquad (\text{II.6.67})$$

When the Fermi energy is close to the bottom or the top of the band, the sine functions cross zero and $|\nabla E| \approx \sqrt{(E_1 a^2 k_x)^2 + (E_2 b^2 k_y)^2 + (E_3 c^2 k_z)^2}$. Inserting this into Eq. II.6.61 leads to a simple integral, which can be performed by introducing new variables $\kappa_x = \sqrt{E_1} a k_x$, and so on for $y$ and $z$. Note that in these coordinates the energy is approximately $E \approx 1/2(\kappa_x^2 + \kappa_y^2 + \kappa_z^2) + \text{const} = \frac{1}{2}\kappa^2 + \text{const}$; the constant energy cut is a sphere, and the surface area of that sphere is $4\pi\kappa^2$. Therefore

$$g = (E_1 E_2 E_3 a^2 b^2 c^2)^{1/3} \frac{1}{\pi^2} \sqrt{2E}. \qquad (\text{II.6.68})$$

The effective mass is defined as

$$M_{ij} = \left( \frac{1}{\hbar^2} \frac{\partial^2 E}{\partial k_i \partial k_j} \right)^{-1} . \tag{II.6.69}$$

As in Problem 3.14, the effective mass tensor is diagonal, and the determinant is found to be

$$\det M = \frac{\hbar^6}{E_1 E_2 E_3 a^2 b^2 c^2} . \tag{II.6.70}$$

The specific heat effective mass is deduced by inserting this into Eq. II.6.68 and comparing the result to the free-electron expression, Eq. II.6.63. Since the total number of electrons is calculated from the density of states (Eq. II.6.62), once the density of state matches, the correspondence between the two models is complete.

For the top of the band the calculation is similar, except the energy should be measured downward from the top of the band. To match the particle numbers expressed by Eq. II.6.62, one has to compare holes in the tight-binding band to the electrons in the nearly free-electron band.

The expression to calculate the specific heat effective mass works for a general band structure, as long as the band filling is close to an extremum in the band structure, and the band can be represented by a quadratic expansion in $k$. No simple formula exists for arbitrary band filling.

## 6.15 Solution: Quantum Hall Effect

The one-electron problem is discussed in great detail by Ziman [3] pp. 313–318.

(a) The energy levels are quantized to the *Landau levels*.

$$E_n = \hbar\omega_c \left( \nu + \frac{1}{2} \right), \quad \nu = 0, 1, 2, \ldots, \tag{II.6.71}$$

where $\omega_c = \frac{eB}{mc}$ is the cyclotron frequency.

(b) The degeneracy of the $n$th Landau level is given by

$$\eta = 2\frac{eBA}{hc} = \frac{\phi}{\phi_0} . \tag{II.6.72}$$

Here the magnetic flux passing through the sample of area A is given by $\phi$, and $\phi_0$ is the magnetic flux quantum.

(c) The density of states of this system is given by a series of energy levels (the Landau levels) with $\hbar\omega_c$ spacing. Assume the magnetic field is large and there are $N = nA$ electrons. As long as $\eta > n$, all the electrons will be in the lowest, $\nu = 0$, level. And if the temperature is small ($k_B T \ll \hbar\omega_c$) the energy of this level is the chemical potential. This chemical potential varies with applied field as

$$\mu = \hbar w_c \frac{1}{2} = \hbar \frac{e}{mc} \frac{B}{2} \; . \tag{II.6.73}$$

When $B$ is reduced from high values, $\eta$ decreases. At $B_0 = n\phi_0$ the lowest Landau level level becomes full, $\eta = N$. At this point electrons begin to move over to the next level and the chemical potential jumps from $\mu = \mu_0/2$ to $\mu' = \frac{3}{2}\mu_0$ (here $\mu_0 = \hbar\frac{e}{mc}n\phi_0 = \pi n\frac{\hbar^2}{m}$). $\mu$ then decreases with decreasing field until the field reaches $B_1 = B_0/2$. Jumps in $\mu$ occur each time the magnetic field crosses $B_i = B_0/i$, where $i$ is an integer, as shown in Figure II.6.4.

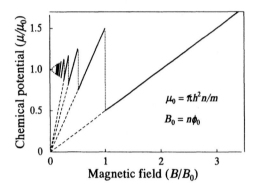

**Fig. II.6.4.** The chemical potential as a function of magnetic field in a two-dimensional electron system.

As the magnetic field goes to zero, the chemical potential approaches $\mu = \mu_0$, as it should for a two-dimensional system in zero magnetic field. At nonzero temperatures (but still where $k_B T \ll \hbar w_c$) the picture is similar, but the jumps will not be as sharp.

(d) Using the numbers given in the problem, we obtain

$$
\begin{aligned}
n \;&=\; 10^{18} \text{ cm}^{-3} \cdot d = 10^{18} \text{ cm}^{-3} \cdot 100 \times 10^{-8} \text{ cm} = 10^{12} \text{ cm}^{-2} \;, \\
B_0 \;&=\; n\phi_0 = 10^{12} \cdot 2.07 \times 10^{-7} \text{ Gauss} = 2 \times 10^5 \text{ Gauss} \\
&=\; 20 \text{ Tesla}, \tag{II.6.74} \\
T \;&\ll\; \frac{\hbar w_c}{k_B} = 1.34 \times 10^{-4} \cdot 2 \times 10^5 \text{ K} = 27 \text{ K} \;.
\end{aligned}
$$

The temperature of liquid He (4.2 K) is sufficiently low for the study of the integer quantum Hall effect.

# 7 Transport Properties

## 7.1 Solution: Temperature Dependent Resistance

Figure II.7.1 shows the Pt resistance data and the Bloch–Grüneisen (BG) function, Eq. I.7.22. The scale for the Pt resistivity was shifted by the residual resistivity (about 1.7 $\Omega$) and adjusted so that the data fits to the theory at 300 K. The BG formula works reasonably well for Pt.

A qualitative understanding of the temperature-dependent resistivity is based on the argument that each phonon scatters the electrons, and therefore the resistivity should be proportional to the number of phonons, $N_{ph}$. In Problem 6.2 $N_{ph}$ was calculated; the result of that calculation is also reproduced in the Figure. Although the linear temperature dependence is reproduced quite well, the number of phonons does not quite match the resistance. At low temperatures $N_{ph} \sim T^2$, the BG formula gives $\rho \sim T^5$, and the experimental data is somewhere in between. At high temperatures $N_{ph}$ stays consistent below $\rho$, whereas the BG formula fits the data well.

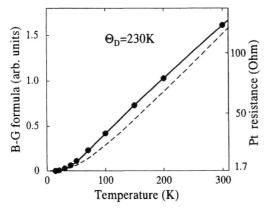

**Fig. II.7.1.** Resistance of a typical Pt resistor (full circles, right scale), the Bloch–Grüneisen formula (full line, left scale), and the number of phonons in the Debye approximation (dashed line, scaled to match the slope of $R(T)$).

## 7.2 Solution: Conductivity Tensor

In terms of components we have

$$j_i \sum_j \sigma_{ij} E_j \equiv \sigma_{ij} E_j . \tag{II.7.1}$$

Here $i$ and $i$ run over $x$, $y$, and $z$ and we introduced the common shorthand notation of summation over indices appearing in pairs (like $j$ in this equation).

Vectors have, by definition, well-defined transformation properties when the reference frame is changed. The transformation can be described by a matrix, $U_{ij}$, such that the components in the new frame of reference $(v_i')$ are related to the components from the old frame of reference $(v_j)$ by

$$v_i' = U_{ij} v_j , \tag{II.7.2}$$

For example, the rotation around the $z$ axis by 90° is described by

$$U_{90}^{(z)} = \begin{bmatrix} 0 & -1 & 0 \\ 1 & 0 & 0 \\ 0 & 0 & 1 \end{bmatrix} . \tag{II.7.3}$$

The rotation around $z$ by 180° is represented by

$$U_{180}^{(z)} = \begin{bmatrix} -1 & 0 & 0 \\ 0 & -1 & 0 \\ 0 & 0 & 1 \end{bmatrix} . \tag{II.7.4}$$

Similar matrices can be constructed for rotations around the $x$ and $y$ axes.

Since both the current and the electric field are vectors, the conductivity must transform according to

$$\sigma_{jk}' = U_{jp} \sigma_{pq} U_{qk}^{-1} , \tag{II.7.5}$$

where $U^{-1}$ is the inverse of $U$; therefore, $U^{-1}U = 1$. (For rotational transformations one can show that $U_{pq}^{-1} = U_{qp}$.) For a crystal of a particular symmetry, the conductivity tensor should not change if a crystal symmetry operation is performed: $\sigma_{ij}' = \sigma_{ij}$ for symmetry operations.

Now look a 90° rotation of the conductivity tensor around the $z$-axis:

$$
\begin{aligned}
\hat{\sigma}_{90}^{(z)} &= U_{90} \hat{\sigma} U_{90}^{-1} \\
&= \begin{bmatrix} 0 & -1 & 0 \\ 1 & 0 & 0 \\ 0 & 0 & 1 \end{bmatrix} \begin{bmatrix} \sigma_{xx} & \sigma_{xy} & \sigma_{xz} \\ \sigma_{yx} & \sigma_{yy} & \sigma_{yz} \\ \sigma_{zx} & \sigma_{zy} & \sigma_{zz} \end{bmatrix} \begin{bmatrix} 0 & 1 & 0 \\ -1 & 0 & 0 \\ 0 & 0 & 1 \end{bmatrix} \\
&= \begin{bmatrix} 0 & -1 & 0 \\ 1 & 0 & 0 \\ 0 & 0 & 1 \end{bmatrix} \begin{bmatrix} -\sigma_{xy} & \sigma_{xx} & \sigma_{xz} \\ -\sigma_{yy} & \sigma_{yx} & \sigma_{yz} \\ \sigma_{zx} & \sigma_{zy} & \sigma_{zz} \end{bmatrix} \\
&= \begin{bmatrix} \sigma_{yy} & -\sigma_{yx} & \sigma_{xz} \\ -\sigma_{xy} & \sigma_{xx} & \sigma_{yz} \\ \sigma_{zx} & \sigma_{zy} & \sigma_{zz} \end{bmatrix} .
\end{aligned} \tag{II.7.6}
$$

Since we know that this rotation corresponds to a symmetry operation for tetragonal crystals, we have $\hat{\sigma} = \hat{\sigma}_{90}^{(z)}$. Comparing each component, we obtain the following relationships:

$$\sigma_{xx} = \sigma_{yy}, \qquad \sigma_{xy} = -\sigma_{yx} . \tag{II.7.7}$$

Next look at a $180°$ rotation about $x$ (which is also a symmetry operation):

$$\hat{\sigma}_{180}^{(x)} = \begin{bmatrix} 1 & 0 & 0 \\ 0 & -1 & 0 \\ 0 & 0 & -1 \end{bmatrix} \hat{\sigma} \begin{bmatrix} 1 & 0 & 0 \\ 0 & -1 & 0 \\ 0 & 0 & -1 \end{bmatrix}$$

$$= \begin{bmatrix} 1 & 0 & 0 \\ 0 & -1 & 0 \\ 0 & 0 & -1 \end{bmatrix} \begin{bmatrix} \sigma_{xx} & -\sigma_{xy} & -\sigma_{xz} \\ \sigma_{yx} & -\sigma_{yy} & -\sigma_{yz} \\ \sigma_{zx} & -\sigma_{zy} & -\sigma_{zz} \end{bmatrix} \tag{II.7.8}$$

$$= \begin{bmatrix} \sigma_{xx} & -\sigma_{xy} & -\sigma_{xz} \\ -\sigma_{yx} & \sigma_{yy} & \sigma_{yz} \\ -\sigma_{zx} & \sigma_{zy} & \sigma_{zz} \end{bmatrix} .$$

Thus from $\hat{\sigma}_{180}^{(x)} = \hat{\sigma}$, we obtain

$$\sigma_{xy} = -\sigma_{xy}, \qquad \sigma_{xz} = -\sigma_{xz} . \tag{II.7.9}$$

Similarly, $\hat{\sigma}_{180}^{(z)} = \hat{\sigma}$ results in $\sigma_{yz} = -\sigma_{yz}$. Putting these results together, we have obtained that

$$\sigma_{xx} = \sigma_{yy} \text{ and } \sigma_{xy} = \sigma_{yz} = \sigma_{xz} = 0 , \tag{II.7.10}$$

resulting in a conductivity tensor which is isotropic in the $x$–$y$ plane:

$$\hat{\sigma} = \begin{bmatrix} \sigma_{xx} & 0 & 0 \\ 0 & \sigma_{xx} & 0 \\ 0 & 0 & \sigma_{zz} \end{bmatrix} . \tag{II.7.11}$$

Note that that three-fold and six-fold rotational symmetries also lead to isotropic $2 \times 2$ conductivity tensor.

## 7.3 Solution: Montgomery Method

We are using the notation of Montgomery [23]; in particular, we will denote the physical dimensions of the sample ($d_1$, $d_2$, and $d_3$) by $l'$. For a thin, square-shaped sample we have $l_1' = l_2' \ll l_3'$. From $R_2/R_1 = 2$ and from Figure 3 of Ref. [23] we get $l_2/l_1 = 1.2$. For a square-shaped sample $l_2'/l_1' = 1$, and, according to Eq. 4 of Montgomery, this corresponds to $\sqrt{\rho_2/\rho_1} = 1.2$. Next, the factor $H$ is determined from Figure 1 by using $l_2/l_1 = 1.2$. We get $H = 9$. Equation 5 results in $\sqrt{\rho_1/\rho_2} = H l_3' R_1 = 0.09$ $\Omega$-cm. Combining these two results yields $\rho_1 = 75$ m$\Omega$-cm and $\rho_2 = 110$ m$\Omega$-cm.

Assuming a given accuracy for the measured resistances and some uncertainty for the readout from the figures, one can analyze the error propagation. We leave this (nontrivial) job to the reader.

For samples of large anisotropy the square configuration does not work, since the resistance anisotropy increases approximately exponentially with $\rho_2/\rho_1$. The solution to this difficulty is to cut the sample shorter in the direction of high resistivity.

## 7.4 Solution: Anisotropic Layer

The resistivity of an anisotropic material is described by a tensor $\hat{\rho}$. The electric field $E$ and the current density $j$ are related by

$$E = \hat{\rho} j \ . \tag{II.7.12}$$

We use a two-dimensional representation, since it matches the experimental conditions:

$$\begin{pmatrix} E_x \\ E_y \end{pmatrix} = \begin{pmatrix} \rho_{xx} & \rho_{xy} \\ \rho_{yx} & \rho_{yy} \end{pmatrix} \begin{pmatrix} j_x \\ j_y \end{pmatrix} \ . \tag{II.7.13}$$

The tensor $\hat{\rho}$ is symmetric ($\rho_{xy} = \rho_{yx}$) and positive definite, therefore the the determinant $\det |\rho_{ij}| = \rho_{xx}\rho_{yy} - \rho_{xy}^2 > 0$.[1]

For long thin samples the voltage across the sample is proportional to the electric field along the long axis, $U = \ell E$, where $\ell$ is the length of the specimen. The current is $I = jA$, where $A$ is the cross section. Therefore, the measured resistance is $R = (\ell/A)(E/j)$. It is important to emphasize that in an anisotropic sample the electric field vector will not necessarily point along the direction of the current flow. In other words, if a voltage is applied along the sample and the current flow starts along the long axis, a nonzero voltage may develop across the sample. Furthermore, in the area where the current is injected into the sample the current density vector may have a quite complicated spatial dependence; that is why we want to consider a long, thin sample in this problem.

In a real measurement the best way to address this and other technical problems (like the nonzero resistance between the electrical contacts and the sample) is to adopt a "four-probe" configuration. In its simplest form a four-probe measurement means that two extra leads are attached to the long, thin specimen so that the four leads divide the sample into approximately equal-length segments. The inner leads serve for the measurement of the voltage;

---

[1] *Proof:* The determinant of a tensor does not depend on the system of reference. In the reference frame where the $\hat{\rho}$ is diagonal, the determinant is $D = \rho_{xx}\rho_{yy}\rho_{zz}$ and the power dissipation is $P = \rho_{xx}j_x^2 + \rho_{yy}j_y^2 + \rho_{zz}j_z^2$. If the determinant was negative, then at least one of the $\rho_{qq}$'s would be negative. In that case one could set the electric field in the other two directions equal to zero to obtain negative power dissipation.

the outer leads are used to inject the current. The "Montgomery method" (see Problem 7.3) is a more sophisticated four-probe arrangement, and it also involves the permutation of voltage and current leads.

To determine the appropriate $E/j$ ratios for the four cuts, let us select a system of reference such that $x$ points along the long axis of cut 2 and $y$ is along cut 4. For cuts 1, 2, 3, and 4 the current density vectors are $(-j/\sqrt{2}, j/\sqrt{2})$, $(0, j)$, $(j/\sqrt{2}, j/\sqrt{2})$, and $(j, 0)$, respectively. We want to determine the electric field along the samples. The field may point in a direction different from the current, therefore the best way to proceed is to calculate the electric field vector and take a scalar product with a unit vector pointing in the direction of the sample ( $(-1/\sqrt{2}, 1/\sqrt{2})$, $(0, 1)$, $(1/\sqrt{2}, 1/\sqrt{2})$, and $(1, 0)$ for the four samples, respectively). For cut 1, we obtain

$$E_x = j(-\rho_{xx} + \rho_{xy})/\sqrt{2}$$
$$E_y = j(-\rho_{xy} + \rho_{yy})/\sqrt{2} \ . \tag{II.7.14}$$

Therefore,

$$E = \left(-\frac{1}{\sqrt{2}}, \frac{1}{\sqrt{2}}\right)\left(\begin{array}{c} j(-\rho_{xx} + \rho_{xy})/\sqrt{2} \\ j(-\rho_{xy} + \rho_{yy})/\sqrt{2} \end{array}\right) \tag{II.7.15}$$

$$= \frac{j}{2}(\rho_{xx} + \rho_{yy} - 2\rho_{xy}) \ . \tag{II.7.16}$$

Similarly, for cut 3:

$$E = \left(\frac{1}{\sqrt{2}}, \frac{1}{\sqrt{2}}\right)\left(\begin{array}{c} j(\rho_{xx} + \rho_{xy})/\sqrt{2} \\ j(\rho_{xy} + \rho_{yy})/\sqrt{2} \end{array}\right) \tag{II.7.17}$$

$$= \frac{j}{2}(\rho_{xx} + \rho_{yy} + 2\rho_{xy}) \ . \tag{II.7.18}$$

Since the measured resistance of these two cuts are equal, we can conclude that $\rho_{xy} = 0$. Doing similar calculations for cut 2 we get $E = j\rho_{yy}$. Using $R_1 = R_3 = R$ and $R_2 = R/2$, we obtain $\rho_{yy} = (1/2)(1/2)(\rho_{xx} + \rho_{yy})$, and therefore $\rho_{xx} = 3\rho_{yy}$.

Finally we look at cut 4. The electric field is $E = j\rho_{xx}$. With the known geometrical factors this leads to the resistance of $R_4 = 3/2R$.

*Challenge*: What would you tell someone who describes an experiment where $R_1 = R_3 = R$ and $R_2 = 3R$?

# 7.5 Solution: Two-Charge-Carrier Drude Model

In the presence of a magnetic field along the $z$-axis, the electric field is given by $\boldsymbol{E} = \hat{\rho}\boldsymbol{j}$, where the resistivity tensor, $\hat{\rho}$, is given by

$$\hat{\rho} = \begin{pmatrix} \rho & -R_H H \\ R_H H & \rho \end{pmatrix} . \tag{II.7.19}$$

$R_H$ is the Hall coefficient.

A single carrier in the Drude model has $\rho = \frac{1}{\sigma} = \frac{m}{ne^2\tau}$ and $R_H = \frac{1}{nec}$. For two carriers the $\boldsymbol{E}$-field remains the same and the current densities add together. Therefore,

$$\hat{\rho}_1^{-1} \boldsymbol{E} + \hat{\rho}_2^{-1} \boldsymbol{E} = \boldsymbol{j} . \tag{II.7.20}$$

The components of $\hat{\rho}_1$ are $\rho_1 = \frac{1}{ne^2\mu_1}$ and $R_{H1} = \frac{1}{nec}$. The components of $\hat{\rho}_2$ are $\rho_2 = \frac{1}{ne^2\mu_2}$ and $R_{H2} = \frac{-1}{nec}$. On the other hand, we know that the total "two-carrier" resistivity tensor, $\hat{\rho}$, must satisfy

$$\hat{\rho}^{-1} \boldsymbol{E} = \boldsymbol{j} . \tag{II.7.21}$$

Putting Eqs. II.7.20 and II.7.21 together, we obtain

$$\hat{\rho}^{-1} = \hat{\rho}_1^{-1} + \hat{\rho}_2^{-1} . \tag{II.7.22}$$

Now

$$\hat{\rho}_1^{-1} = \begin{pmatrix} \rho_1 & -R_{H1} H \\ R_{H1} H & \rho_1 \end{pmatrix}^{-1} = \frac{\begin{pmatrix} \rho_1 & R_{H1} H \\ -R_{H1} H & \rho_1 \end{pmatrix}}{\begin{vmatrix} \rho_1 & -R_{H1} H \\ R_{H1} H & \rho_1 \end{vmatrix}} . \tag{II.7.23}$$

The determinant in the denominator is simply equal to $\rho_1^2 + R_{H1}^2 H^2$.

(a) Putting all this together to solve for $\Delta\rho$, we obtain the relative magnetoresistance,

$$\Delta\rho/\rho = \left(\frac{e}{c}\right)^2 \mu_1 \mu_2 H^2 , \tag{II.7.24}$$

where $\rho = \rho_1\rho_2/(\rho_1 + \rho_2)$. By definition, $\mu_1$ and $\mu_2$ are positive, so the magnetoresistance is always positive.

(b) We can now solve for $R_H$:

$$R_H = \frac{1}{nec} \frac{\mu_1 - \mu_2}{\mu_1 + \mu_2} , . \tag{II.7.25}$$

If the mobilities are equal, then the Hall effect is zero. This is expected for a system of perfect electron–hole symmetry.

(c) The semiconductor has a temperature-dependent carrier density given by

$$n = n_0 e^{-\Delta/k_B T} . \tag{II.7.26}$$

This means that now

$$\rho_1 = \frac{1}{ne^2\mu_1} = \frac{e^{\Delta/k_B T}}{n_0 e^2 \mu_1} , \tag{II.7.27}$$

and a similar equation for $\rho_2$. So plugging these into Eq. II.7.24, we obtain a temperature-dependent magnetoresistance given by

$$\Delta\rho = e^{\Delta/k_BT} \frac{1}{n_0c^2} \frac{\mu_1\mu_2}{\mu_1 + \mu_2} H^2 e^{\Delta/k_BT} \tag{II.7.28}$$

The Hall coefficient is similarly solved from Eq. II.7.25:

$$R_H = \frac{1}{n_0ec} \frac{\mu_1 - \mu_2}{\mu_1 + \mu_2} e^{\Delta/k_BT}, \tag{II.7.29}$$

Note that the Hall effect increases as the temperature is lowered and the carrier concentration decreases. However, since the difference $\mu_1 - mu_2$ appears in the result, the temperature dependence of the mobilities also influences the Hall effect.


## 7.6 Solution: Thermal Conductivity

(a) Use the Onsager relations,

$$\text{Electical current}: \quad j_e = L_{11}E - L_{12}\frac{\nabla T}{T} \tag{II.7.30}$$

$$\text{Thermal current}: \quad j_q = L_{21}E - L_{22}\frac{\nabla T}{T} \tag{II.7.31}$$

$$L_{21} = L_{12}, \tag{II.7.32}$$

and solve for $E = 0$. The thermal current becomes

$$j_q = -L_{22}\frac{\nabla T}{T} \quad \Longrightarrow \quad \kappa' = \frac{L_{22}}{T}. \tag{II.7.33}$$

With no electrical current density ($j_e = 0$), we obtain

$$E = \frac{L_{12}}{L_{11}} \frac{\nabla T}{T} \tag{II.7.34}$$

and

$$j_q = \left(\frac{L_{12}^2}{L_{11}} - L_{22}\right) \frac{\nabla T}{T}. \tag{II.7.35}$$

This leads to

$$\kappa = \left(-\frac{L_{12}^2}{L11} + L_{22}\right) \frac{1}{T}, \tag{II.7.36}$$

which together with $\kappa'$ yield

$$\Delta\kappa = \frac{L_{12}^2}{L_{11}} \frac{1}{T}. \tag{II.7.37}$$

The Peltier coefficient is $\Pi = L_{12}/L_{11}$, and the electrical conductivity is $\sigma = L_{11}$. Inserting these into Eq. II.7.37, we obtain

$$\Delta\kappa = \Pi^2\sigma\frac{1}{T} = TS^2\sigma . \qquad (\text{II}.7.38)$$

(b) Typical values of $S$ (or $\Pi$), $\kappa$, and $\sigma$ can be found in textbooks or handbooks on materials properties, like Landolt–Börnstein [21]. Alternatively, we can use the "free-electron" estimates:

$$\frac{\Delta\kappa}{\kappa} = TS^2\frac{\sigma}{\kappa} = \frac{S^2}{L} , \qquad (\text{II}.7.39)$$

where $L \approx \frac{\pi^2 k_B^2}{3e^2}$, and $S \approx \frac{\pi^2}{6}\frac{k_B}{e}\left(\frac{k_B T}{\varepsilon_F}\right)$. Therefore,

$$\frac{\Delta\kappa}{\kappa} \approx \frac{1}{2}\left(\frac{k_B T}{\varepsilon_F}\right)^2 \approx 10^{-4} \qquad \text{at } T = 100 \text{ K}. \qquad (\text{II}.7.40)$$

## 7.7 Hint: Residual Resistivity

Estimate a scattering cross section of conduction electrons scattered by impurities using dimensional analysis, assuming that the strength of the interaction is $Ze^2$, where $Z$ is the extra charge on the impurity (in other words, $Z$ is the deviation of the atomic number of the impurity from that of copper), and $e$ is the electric charge. The only other relevant parameter is the Fermi energy of copper $E_F \approx 7$ eV. Assume that electrons move with the Fermi velocity and calculate the relaxation time for a given concentration of impurities. Use the Drude model to calculate the resistivity.

## 7.8 Solution: Electric and Heat Transport

(a) The typical resistivities for metals and intrinsic semiconductors are sketched in Figure II.7.2. At very low temperatures, the metal has a constant $\rho$ due to impurities. For $T < \Theta_D$, we have $\rho \sim \text{constant} + T^\alpha$, where the exponent $\alpha$ ranges between 2 and 5. $\rho$ is proportional to $T$ at high temperatures, since the resistivity is due to electron–phonon scattering, and the number of phonons is proportional to $T$.[2] Good metals have resistivities in the range of a few $\mu\Omega$-cm.

The resistivity of a semiconductor is approximately given by

$$\rho = \rho_0 e^{\Delta/k_B T} , \qquad (\text{II}.7.41)$$

which comes from the number of carriers thermally activated across the energy gap, $\Delta$.

---

[2] See also Problem 7.1.

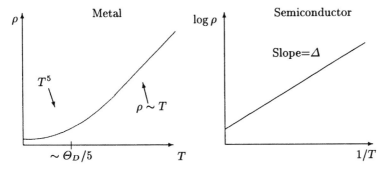

**Fig. II.7.2.** Resistivity as a function of temperature for a typical metal and semiconductor. Note the inverse temperature and log resistivity scales for the semiconductor.

(b) The temperature dependence of the thermal conductivity of metals and insulators looks similar, but the physical processes are different. In pure metals the thermal conductivity is dominated by electrons. At low temperatures it is proportional to the temperature; at high $T$ it is approximately constant:

$$\kappa \sim T/\rho_0 \quad \text{or} \quad \kappa \sim \text{const.} . \tag{II.7.42}$$

Pure metals exhibit a large peak in the thermal conductivity.[3] The Wiedemann–Franz ratio, $\alpha = \kappa\rho/T$ remains approximately constant over the whole range of temperatures. In alloys, where the electrical resistivity is dominated by impurity scattering, the thermal conductivity increases smoothly.[4]

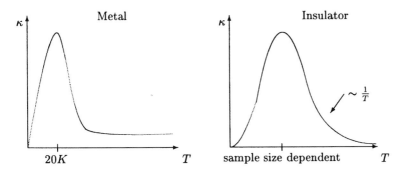

**Fig. II.7.3.** Thermal conductivity as a function of temperature for a typical metal and semiconductor.

---

[3] For gold, see G.K. White, *Proc. Phys. Soc. (London)* **A247**, 441 (1953); for copper see R. Berman and D.K.C. McDonald, *Proc. Roy. Soc. (London)* **A211**, 122, (1952)

[4] Measurements on an extensively used standard of thermal conductivity, the stainless steel SRM735, are described by J.G. Hust and P.J. Giarratano, NBS Special Publications 260-46 (1975) and discussed by R.S. Grave, R.K. Williams and J.P. Moore in *"Thermal Conductivity 16"* Ed. D.C. Larsen, p. 343 (Plenum, 1983).

The thermal conductivity of insulators is due to phonons. At high temperature it is much smaller than the thermal conductivity of metals, and it follows an approximate $1/T$ behavior, since the mean free path of the phonons decreases with temperature (See Eq. I.7.9). At lower temperatures the mean free path usually reaches the sample size and the thermal conductivity starts to drop. The crossover temperature depends on the size of the sample. When the temperature approaches zero the thermal conductivity drops rapidly, as the number of high energy phonons decreases.

For detailed discussions, see Ashcroft and Mermin [1] pp. 495–505, Kittel [2] pp. 144–149, Ibach and Lüth [4] pp. 76–80, or Ziman [3] 239–244.

## 7.9 Solution: Conductivity of Tight-Binding Band

(a) For a two-dimensional square lattice the conductivity is isotropic, $\sigma_{xx} = \sigma_{yy}$. Therefore Eq. I.7.23 becomes

$$\sigma_{xx} = \frac{1}{2\pi} \frac{\tau e^2}{h} \oint dk |v| , \qquad \text{(II.7.43)}$$

where the integral is carried out along the Fermi line.

The velocity of the electrons, $v$, is calculated using Eq. I.3.12:

$$v = \frac{1}{\hbar} \left( \begin{array}{c} \partial E/\partial k_x \\ \partial E/\partial k_y \end{array} \right) . \qquad \text{(II.7.44)}$$

With the energy given in Eq. I.7.26, we obtain

$$v_x = \frac{E_1 a}{\hbar} \sin k_x a \qquad \text{(II.7.45)}$$

$$v_y = \frac{E_1 a}{\hbar} \sin k_y a . \qquad \text{(II.7.46)}$$

According to the solution of the Boltzmann equation, we are interested in the electron velocity at the Fermi surface (in two dimensions, the Fermi line) only. Figure II.7.4 illustrates the Fermi line for a nearly-half filled and half-filled tight-binding band. Since $k_x$ and $k_y$ in Eq. II.7.46 must be on the Fermi line,

$$|v| = (v_x^2 + v_y^2)^{1/2} = \frac{E_1 a}{\hbar} \sqrt{\sin^2 k_x a + \sin^2 k_y a} . \qquad \text{(II.7.47)}$$

Putting this into Eq. II.7.43, we obtain the expression

$$\sigma_{xx} = \frac{1}{2\pi} \frac{e^2}{h} 4 \int_0^{\pi/a} \sqrt{2} \tau \frac{E_1 a}{\hbar} \sqrt{\sin^2 k_x a + \sin^2 \left( \frac{\pi}{a} - k_x \right) a} \, a \, dk_x , \qquad \text{(II.7.48)}$$

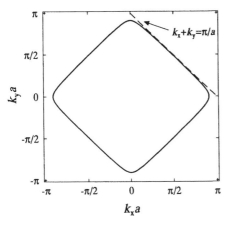

Fig. II.7.4. The Fermi line for a two−dimensional tight−binding model. The solid line corresponds to a nearly half-filled band. A segment of the Fermi line for a half-filled band is indicated by the dashed line.

where the factor 4 in front of the integral is to account for all four sides of the Fermi line, and we have used the equation of the line given in Figure II.7.4, $k_y = \frac{\pi}{a} - k_x$. This expression can be simplified to

$$\sigma_{xx} = \frac{2\sqrt{2}\,e^2}{\pi}\frac{E_1 a}{h}\,\tau\frac{}{\hbar}\int_0^{\pi/a} \sqrt{2}\sin k_x a\ dk_x \qquad (\text{II.7.49})$$

$$= \frac{4\,e^2}{\pi\ h}\tau\frac{E_1}{\hbar}\ . \qquad (\text{II.7.50})$$

Note that $\frac{e^2}{h} = [6.5\ \text{k}\Omega]^{-1}$, and the dimensionless ratio $\frac{4E_1}{\hbar/\tau}$ relates the bandwidth $(4E_1)$ to the relaxation rate $(1/\tau)$.

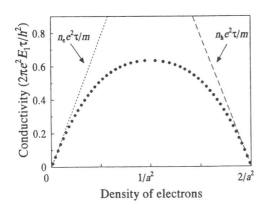

Fig. II.7.5. The conductivity as a function of electron density. The dots are a result of the numerical integration of Eq. II.7.43. Note that the Drude expression (Eq. I.7.7) is valid for nearly empty or nearly full bands only.

(b) The Drude conductivity is given by

$$\sigma_{xx} = \frac{ne^2\tau}{m}\ , \qquad (\text{II.7.51})$$

where $m$ is the effective mass and $n = \frac{1}{a^2}$ (a half-filled band means one electron per site). This can be made identical to the result for part (a) if we define the effective mass to be

$$m = \frac{\pi^2}{2} \frac{\hbar^2}{E_1 a^2} .$$

(II.7.52)

Note that the effective mass at the bottom of the energy band is given by

$$E = E_0 + E_1 \frac{1}{2}(k_x^2 + k_y^2)a^2 = E_0 + \frac{1}{2} \frac{\hbar^2 k^2}{M}$$

(II.7.53)

$$M = \frac{\hbar^2}{E_1 a^2} .$$

(II.7.54)

$M$ is quite different from the effective mass we determined above. The discrepancy illustrates that the concept of effective mass works well only for nearly empty bands (or, by introducing holes instead of electrons, for nearly full bands). This is also apparent in Figure II.7.5, where the results of the numerical integration of Eq. II.7.43 are plotted for various band fillings.

## 7.10 Solution: Hall Effect in Two-Dimensional Metals

N.P. Ong [22] provides a geometrical interpretation for the solution to the Boltzmann equation (I.7.24):

$$\sigma_{xy} = \frac{e^2}{h} \frac{2\phi}{\phi_0} ,$$

(II.7.55)

where $\phi$ is the magnetic flux within the area swept by the mean free path vector, $\mathbf{l} = \tau \mathbf{v}_F$, and $\phi_0 = hc/e = 4 \times 10^{-7}$ Gauss-cm$^2$ is the magnetic flux quantum.[3] It is convenient to rewrite the conductivity (see Eq. I.7.23) in similar terms:

$$\sigma_{xx} = \frac{1}{2\pi} \frac{e^2}{h} \oint dk|\mathbf{l}| ,$$

(II.7.56)

where the integral is carried out along the Fermi line.

Once the components of the Hall conductivity tensor are determined, we can obtain the resistivity and the Hall coefficient from Eqs. I.7.4 and I.7.5. For "normal" magnetic fields, when $\rho \gg R_H H$, we obtain

$$\rho \approx \frac{1}{\sigma_{xx}} \quad \text{and} \quad R_H \approx -\frac{\sigma_{xy}}{H\sigma_{xx}^2} .$$

(II.7.57)

To calculate $\phi$ in Eq. II.7.55 we need to plot the endpoints of the Fermi velocity vectors. Assume that the band is close to half-filled and the Fermi line looks like the one plotted in Figure II.7.4. The calculation of the Fermi

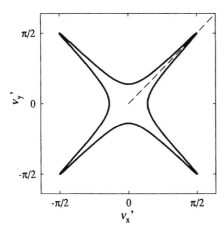

**Fig. II.7.6.** The solid line corresponds to the end points of the normalized Fermi velocity, $v' = v(k_F)E_1 a/h$, as the wavevector $k_F$ scans the Fermi surface of a nearly half-filled band. The dashed line indicates a portion of the same plot for a half-filled band.

velocity is discussed in the solution to Problem 7.9. As the $k$-vector scans the Fermi line, the end points of the $v(k)$ vectors make up a line shown in Figure II.7.6. The area within this curve, $A$, can be determined by numerical integration for any band-filling. As long as $\tau$ is independent of the wavevector, the flux is $\phi = H\tau^2 A$.

For a half-filled band the curve in Figure II.7.6 collapses to four straight lines starting from the origin; one of these lines is shown in the Figure as a dashed line. Therefore the total area of the curve is zero for a half-filled band. From Eq. II.7.55, we obtain

$$\sigma_{xy} = 0. \tag{II.7.58}$$

The result, combined with Eqs. II.7.50 and II.7.57, tells us that the Hall coefficient is $R_H = 0$. This is a very general result: The Hall effect is zero if there is electron–hole symmetry in the system. When the band filling is different from $1/2$, we will have a finite Hall coefficient; $R_H > 0$ for a slightly more than half-filled band. By numerical integration one can easily obtain the transport coefficients at arbitrary electron density. Figure II.7.7 illustrates the Hall coefficient as a function of band filling. The Drude result proves to be totally invalid in the neighborhood of half-filling: in this regime, instead of following $R_H \sim 1/n$, the Hall coefficient is actually proportional to $n$.

## 7.11 Solution: Free-Electron Results from the Boltzmann Equations

Free-electrons have an energy of

$$E = \frac{\hbar^2 k^2}{2m}, \tag{II.7.59}$$

---

[3] In this formula, $\phi_0$ is defined with $1e$ instead of $2e$ as is used when discussing superconductivity.

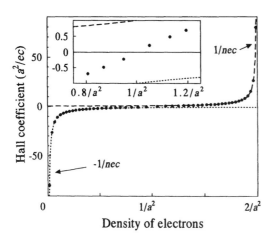

**Fig. II.7.7.** Hall coefficient as a function of the electron density. The inset shows the behavior in the neighborhood of a half-filled band. The dots are results of numerical integration, and the dashed lines represent the Drude result, Eq. I.7.8.

and therefore they have a velocity of

$$v_i = \frac{\hbar k_i}{m} . \tag{II.7.60}$$

Also, the conductivity, $\sigma$, is isotropic. Combining this with the solution to the Boltzmann transport equation as given in this problem, we obtain

$$\sigma = \frac{1}{3}(\sigma_{xx} + \sigma_{yy} + \sigma_{zz}) = \frac{1}{3}\frac{1}{4\pi^3}\frac{e^2\tau}{\hbar}\int\frac{\frac{\hbar^2 k^2}{m^2}dS_F}{\frac{\hbar|k|}{m}} . \tag{II.7.61}$$

Integration over the Fermi surface just results in the area of the Fermi surface, $4\pi^2 k_F^2$. Therefore,

$$\sigma = \frac{1}{3\pi}\frac{e^2\tau}{m}k_F^3 . \tag{II.7.62}$$

We can express $k_F$ in terms of $n$ since we know that the total number of electrons, $N$, is within the volume of the Fermi surface sphere:

$$N = 2\frac{4\pi}{3}k_F^3\left(\frac{L}{2\pi}\right)^3 , \tag{II.7.63}$$

where the factor of two is for spin, and $n = N/L^3$. Therefore,

$$n = \frac{1}{3\pi}k_F^2 , \tag{II.7.64}$$

and finally

$$\sigma = \frac{ne^2\tau}{m} . \tag{II.7.65}$$

## 7.12 Solution: $p$–$n$ Junctions

(a) The total current at a given bias voltage, $V$, is written

$$I = I_{pn}^e + I_{np}^e e^{eV/k_{\mathrm{B}}T} + I_{pn}^h e^{eV/k_{\mathrm{B}}T} + I_{np}^h . \tag{II.7.66}$$

Using the relations in Eq. I.7.30, the total current can be rewritten

$$I = I_{pn}^e \left[1 - e^{eV/k_{\mathrm{B}}T}\right] + I_{np}^h \left[1 - e^{eV/k_{\mathrm{B}}T}\right] . \tag{II.7.67}$$

These currents add up, and result in the $I$–$V$ curve plotted in Figure II.7.8. This device is a standard diode.

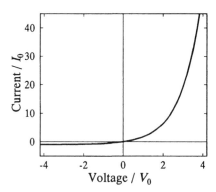

**Fig. II.7.8.** Current–voltage relationship for a $p$–$n$ diode junction, according to Eq. II.7.67. The voltage scale is set by the operating temperature, $V = k_{\mathrm{B}}T/e$. The current scale depends on the zero bias diffusion currents of electrons and holes, $I_0 = I_{pn}^e + I_{np}^h$.

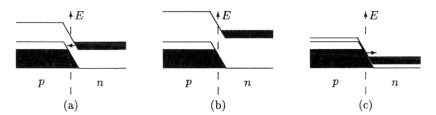

**Fig. II.7.9** Energy level diagrams for various bias voltages for a tunnel diode.

(b) When tunneling is allowed, the device becomes a tunnel diode. We can pictorially describe the possible bias voltage situations in Figure II.7.9. Figure II.7.9(a) shows the energy levels when a small bias voltage is applied and tunneling is allowed. Figure II.7.9(b) shows a large forward bias; tunneling is not allowed and the device now acts as a regular diode. In Figure II.7.9(c) the reverse biased junction is shown; we see tunneling is allowed at all reverse bias voltages.

The energy level diagrams in Figure II.7.9 can be translated into the $I$–$V$ curves of Figure II.7.10. The current–voltage characteristics of the tunnel diode are made up of a tunneling part and a conventional diode part.

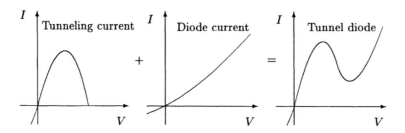

**Fig. II.7.10.** The current–voltage relations for a tunnel diode are made up of a tunneling portion and a conventional diode portion.

A nice discussion of diodes and tunnel diodes is given on pp. 28–42 of Dunlap [7].

# 8 Optical Properties

## 8.1 Solution: Fourier Transform Infrared Spectroscopy

(a) Let us assume that the sample transmits only at a single frequency $\omega$: $T(\omega) \propto \delta(\omega' - \omega)$. In a Michelson interferometer, for monochromatic radiation, the measured intensity will behave as $\sim\cos^2(2\pi \; 2x/\lambda)$, where $\lambda = 2\pi c/\omega'$, $x$ is the displacement of the moving mirror, and the extra factor of 2 is necessary to account for the light going back and forth to the mirror. Accordingly, the intensity will oscillate as $\frac{1}{2}\cos(4\omega'x/c) + 1/2$. If we choose the constant in Eq. I.8.26 to be $\alpha = 4/c$, then the result of integration will be indeed $T \propto \delta(\omega' - \omega)$. The nonoscillating term in the intensity will lead to a "false" peak at $\omega = 0$. An arbitrary transmission function can be viewed as the superposition of many $\delta(\omega' - \omega)$ functions.

(b) The frequency resolution of the instrument depends on the path length $L$ of the mirror. Mathematically, the path length corresponds to a finite cutoff in the integral Eq. I.8.26, resulting in all kinds of unwanted, artificial low-frequency oscillations in the Fourier spectrum, below the frequency $\Omega \sim \alpha/2\pi L$. These false signals can be "filtered out" (usually by digital processing), but only by paying a price: We cannot distinguish between frequencies $\omega$ and $\omega + \Omega$.

A more physical picture is obtained if we assume that the sample transmits at two closely spaced frequencies, $\omega_1$ and $\omega_2$. For zero displacement, every frequency leads to constructive interference. As the mirror moves, the interference pattern due to the two different colors are very similar for small displacements. Significant differences between the two patterns develop only if $x \sim L = \alpha/2\pi(\omega_2 - \omega_1)$.

(c) With the given mirror speed, radiation with frequency $\omega$ is converted to an audio frequency $f = v/\Delta x = \alpha\omega v/2\pi = 2\omega v/c\pi$, where $\Delta x = 2\pi/\alpha\omega$ is the the distance between two maxima in the interference pattern. Accordingly, $\omega_0 = (\pi/2)(c/v)f_0$ and the corresponding wavenumber is $1/\lambda = (1/4)(f/v) = 1000 \text{ cm}^{-1}$.

Infrared spectroscopy is discussed briefly by Ibach and Lüth [4] p. 255.

## 8.2 Solution: Optical Mode of KBr

The frequency of the KBr vibrational mode is $\omega = \sqrt{2\kappa/m}$, where $\kappa = aB$ is the restoring force and $m$ is the reduced mass, $m = m_1 m_2/(m_1 + m_2)$. With atomic numbers 40 and 80 for K and Br, respectively, one obtains $\omega^2 = (1.48 \times 10^{11})(6.59 \times 10^{-8})/26.6 \times 1.67 \times 10^{-24} = 4.1 \times 10^{26}$ sec$^{-2}$. Therefore $\nu = \omega/2\pi = 3.2 \times 10^{12}$ Hz and $1/\lambda = 2 \times 10^{13}/6.28 \times 3 \times 10^{10}$ cm$^{-1} \approx 100$ cm$^{-1}$. The actual measured value of the KBr TO frequency is $3.8 \times 10^{12}$ Hz and the LO mode frequency is $5.0 \times 10^{12}$ Hz, from neutron scattering by A.D.B. Woods et al. [*Phys. Rev. B*, **131**, 1025 (1963)].

## 8.3 Solution: Direct-Gap Semiconductor

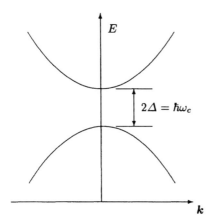

Fig. II.8.1. Conduction and valence bands for a direct-gap semiconductor

We will use Eq. I.8.22 to calculate the conductivity, with Eq. I.8.20 for the joint density of states. A sketch of the valence and conduction bands of the semiconductor described in this problem is in Figure II.8.1. The conduction band energy is described by

$$E_c = E_0 + \Delta + \frac{\hbar^2}{2m_e}(k_x^2 + k_y^2 + k_z^2) , \qquad (II.8.1)$$

and the valence band is described by

$$E_v = E_0 - \Delta - \frac{\hbar^2}{2m_h}(k_x^2 + k_y^2 + k_z^2) . \qquad (II.8.2)$$

The energy difference for direct (i.e., $k$ does not change) transitions is

$$\Delta E = 2\Delta + \frac{\hbar^2}{2} \frac{m_e + m_h}{m_e m_h} (k_x^2 + k_y^2 + k_z^2) = 2\Delta + \frac{\hbar^2}{2\mu} k^2 . \qquad \text{(II.8.3)}$$

We have defined $\mu = \frac{m_e m_h}{m_e + m_h}$.

We can use the relations $d^3k = 4\pi k^2 dk$, $k dk = \frac{\mu}{\hbar^2} dE$ and $\frac{dk^2}{4\pi^3} = g(E)dE$ to obtain the density of states $g$:

$$g(E) = \frac{\sqrt{2}}{\pi^2} \left(\frac{\mu}{\hbar^2}\right)^{3/2} \sqrt{E - E_c} . \qquad \text{(II.8.4)}$$

where $E_c = 2\Delta$. The real part of the conductivity is then calculated to be

$$\sigma_1 = \frac{1}{\sqrt{2\pi}} \frac{e^2 \mu}{\hbar m} f \sqrt{\frac{\mu \omega^2}{\hbar^2}} \sqrt{\omega - \omega_c} . \qquad \text{(II.8.5)}$$

In one dimension, we use $2\frac{dk}{2\pi} = g(E)dE$, and $d\Delta E = \frac{\hbar^2}{\mu} k dk$ to obtain

$$g(E) = \frac{1}{\sqrt{2\pi}} \sqrt{\frac{\mu}{\hbar^2}} \frac{1}{\sqrt{E - E_c}} . \qquad \text{(II.8.6)}$$

At the energy corresponding to the direct gap transition, the optical conductivity exhibits van Hove singularities. Many optical devices, such as solid

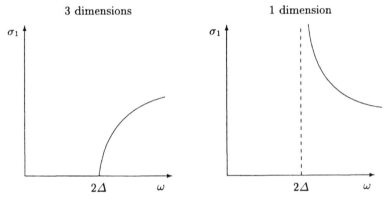

Fig. II.8.2. $\sigma_1(\omega)$ for a one-dimensional and three-dimensional direct-gap semiconductor.

state lasers and photodetectors, involve these transitions.[1] $\sigma_1(\omega)$ is plotted in Figure II.8.2 for the three-dimensional and one-dimensional cases.

---

[1] For a detailed discussion, see Yu and Cardona [6] pp. 252–263.

## 8.4 Solution: Inversion Symmetry

(a) The first class of inversion symmetry points in the diamond lattice are halfway between any nearest-neighbors. In addition, there is another class of inversion symmetry points. To obtain these, one has to move along the nearest-neighbor direction, and continue to the next nearest-neighbor along this line. The inversion symmetry point is halfway to the next nearest-neighbor.

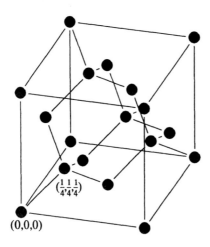

**Fig. II.8.3.** The diamond lattice consists of two interpenetrating $fcc$ lattices which are displaced relative to each other by $\frac{1}{4}$ of the long diagonal of the cube.

Figure II.8.3 shows the diamond lattice in its $fcc$ representation. The two inversion symmetry points are located at $(\frac{1}{8},\frac{1}{8},\frac{1}{8})$, and $(\frac{7}{8},\frac{7}{8},\frac{7}{8})$.

(b) The wavelength of light is typically much longer than the lattice spacing, therefore the light couples to the $k = 0$ mode. At $k = 0$ the atomic displacements are exactly the same for every unit cell and it is sufficient to investigate a single cell. Silicon has diamond structure with two atoms per unit cell. Since the center of mass does not move, the two Si atoms must move exactly opposite to each other. The two atoms are identical and they carry the same charge. There is no net dipole moment associated to this mode.

In a more general sense, the absence of dipole activity for this mode is related to the inversion symmetry of the crystal. One can prove that in the presence of inversion symmetry the vibrational modes at $k = 0$ are either even or odd; i.e., under an inversion the atomic displacements either remain the same (even) or become exactly opposite (odd) during an inversion. Only the odd modes can have a dipole moment and thus have IR activity, and the optical mode of silicon is an even mode.

## 8.5 Solution: Frequency-Dependent Conductivity

(a) For two-dimensional systems the Fermi surface becomes a "Fermi line"; $dS_F \to dl_F$, and Equation I.8.25 becomes

$$\sigma_{ij} = \frac{e^2}{2\pi^2} \int \frac{\tau v_i v_j}{1 - i\tau\omega} \frac{dl_F}{\hbar|v|} . \tag{II.8.7}$$

We neglected the term $K \cdot v$ in the denominator, since $K \cdot v \ll \omega$ for typical electromagnetic waves of wavenumber $K$ and typical electron velocities. Since the square lattice has an isotropic conductivity, we can rewrite this expression as

$$\sigma_{ij} = \frac{1}{2}(\sigma_{xx} + \sigma_{yy}) = \frac{e^2}{2\pi^2\hbar} \int \frac{\tau|v|}{1 - i\tau\omega} dl_F . \tag{II.8.8}$$

Note that the mean free path is $l = \tau|v|$. Including $l$ we obtain

$$\sigma_{ij} = \frac{e^2 l}{4\pi^2\hbar} \int \frac{1}{1 - \frac{il\omega}{|v|}} dl_F . \tag{II.8.9}$$

We then separate this into its real and imaginary parts:

$$\sigma_1 \equiv \mathrm{Re}(\sigma) = A \int \frac{1}{1 - \frac{B^2}{|v|^2}} dl_F \tag{II.8.10}$$

$$\sigma_2 \equiv \mathrm{Im}(\sigma) = A \int \frac{\frac{B}{|v|}}{1 - \frac{B^2}{|v|^2}} dl_F . \tag{II.8.11}$$

where we have defined $A = \frac{e^2 l}{4\pi^2\hbar}$ and $B = l\omega$.

For a half-filled tight-binding band we have $E = E_0 + E_1(\cos k_x a + \cos k_y a)$; therefore we can solve for the velocity:

$$v = \frac{1}{\hbar}\nabla E = \frac{E_1 a}{\hbar} \begin{pmatrix} -\sin k_x a \\ \sin k_y a \end{pmatrix} . \tag{II.8.12}$$

On the Fermi line we know that $|k_x| = |\frac{\pi}{a} - k_y|$ (see Figures II.3.11 and II.3.13). Therefore $|\sin k_x a| = |\sin k_y a|$, and we obtain

$$|v|^2 = (\sin^2 k_x a + \sin^2 k_y a)\left(\frac{E_1 a}{\hbar}\right)^2 = 2\sin^2 k_x a \left(\frac{E_1 a}{\hbar}\right)^2 . \tag{II.8.13}$$

We can also rewrite $dl_F = \sqrt{2}dk_x$. Therefore the real part of the conductivity becomes (remembering a factor of 4 for the four sides of the Fermi line)

$$\sigma_1 = 4A \int_0^{\pi/a} \frac{1}{1 + B^2 \frac{\hbar^2}{2E_1^2 a^2} \frac{1}{\sin^2 k_x a}} \sqrt{2}\, dk_x . \tag{II.8.14}$$

For simplicity we now define $C = B\frac{\hbar}{\sqrt{2E_1a}} = l\omega\frac{\hbar}{\sqrt{2E_1a}}$. This integral can be solved using appropriate variable substitution:

$$
\begin{aligned}
\sigma_1(\omega) &= 4A\pi\left(1 - \frac{C}{\sqrt{C^2+1}}\right)\frac{\sqrt{2}}{a} \\
&= \frac{\sqrt{2}e^2l}{\pi\hbar a}\left(1 - \frac{\omega\tau^*}{\sqrt{(\omega\tau^*)^2+1}}\right), \quad\quad \text{(II.8.15)}
\end{aligned}
$$

where we have defined $\tau^* = \frac{l\hbar}{\sqrt{2E_1a}}$.

We similarly obtain the result for the imaginary part of the conductivity:

$$
\begin{aligned}
\sigma_2(\omega) &= A\frac{4\sqrt{2}}{a}\frac{C}{\sqrt{C^2+1}}\ln\left(\frac{\sqrt{C^2+1}+1}{\sqrt{C^2+1}-1}\right) \quad\quad \text{(II.8.16)} \\
&= \frac{\sqrt{2}e^2l}{\pi^2\hbar a}\frac{\omega\tau^*}{\sqrt{(\omega\tau^*)^2+1}}\ln\left(\frac{\sqrt{(\omega\tau^*)^2+1}+1}{\sqrt{(\omega\tau^*)^2+1}-1}\right). \quad \text{(II.8.17)}
\end{aligned}
$$

(b) The area under $\sigma_1(\omega)$ is

$$
I_1 = \int_0^\infty \sigma_1(\omega)\,d\omega = \frac{\sqrt{2}e^2l}{\pi\hbar a}\int_0^\infty\left(1 - \frac{\omega\tau^*}{\sqrt{(\omega\tau^*)^2+1}}\right)d\omega. \quad \text{(II.8.18)}
$$

Using $x = \omega\tau^*$ and substituting in the definition from part (a) for $\tau^*$, this integral becomes

$$
I_1 = \frac{2}{\pi}\frac{e^2}{\hbar^2}E_1\int_0^\infty\left(1 - \frac{x}{\sqrt{x^2+1}}\right)dx = \frac{2}{\pi}\frac{e^2}{\hbar^2}E_1. \quad \text{(II.8.19)}
$$

In the two-dimensional free-electron case, we have

$$
\sigma = \frac{ne^2\tau}{m}\frac{1}{1+\omega^2\tau^2}, \quad\quad \text{(II.8.20)}
$$

which has an area of

$$
I_{FE} = \int_0^\infty \sigma(\omega)\,d\omega = \frac{\pi}{2}\frac{ne^2}{m}. \quad\quad \text{(II.8.21)}
$$

We can express this in terms of the Fermi energy, $E_F$, using $E_F = \frac{\hbar^2 k_F^2}{2m}$ and $\frac{N}{L^2} = \frac{k_F^2}{2\pi}$. $I_{FE}$ then becomes

$$
I_{FE} = \frac{\pi}{2}\frac{1}{2\pi}\frac{2}{\hbar^2}e^2E_F = \frac{1}{2}\frac{e^2}{\hbar^2}E_F. \quad\quad \text{(II.8.22)}
$$

The tight-binding approximation is only meaningful if $E_1 \ll E_F$. This means that

$$
I_1 \ll I_{FE}. \quad\quad \text{(II.8.23)}
$$

The oscillator sum rule is violated with the missing oscillator strength going into the interband transitions, which were entirely neglected in our tight-binding calculation.

(c) To show that $\sigma_1(\omega)$ and $\sigma_2(\omega)$ from Eqs. II.8.15 and II.8.17, respectively, satisfy the Kramers–Kronig relation, we must calculate

$$\sigma_2(\omega) = \frac{2\omega}{\pi} \int_0^\infty \frac{\sigma_1(\omega')}{\omega'^2 - \omega^2} \, d\omega' \tag{II.8.24}$$

$$= \frac{2\omega}{\pi} \int_0^\infty \frac{\sqrt{2}e^2 l}{\pi \hbar a} \left( 1 - \frac{\omega'\tau^*}{\sqrt{(\omega'\tau^*)^2 + 1}} \right) \frac{d\omega'}{\omega'^2 - \omega^2} . \tag{II.8.25}$$

The two additive parts of this integral can be done independently. Without the constants in front of the integrals, the first part is

$$\int_0^\infty \frac{d\omega'}{\omega'^2 - \omega^2}$$

$$= \frac{1}{2\omega} \left[ \ln\left( \frac{\omega + \omega'}{\omega - \omega'} \right) \Big|_{\omega'=0}^{\omega'=\omega-\varepsilon} + \ln\left( \frac{\omega + \omega'}{\omega - \omega'} \right) \Big|_{\omega'=\omega+\varepsilon}^{\omega'=\infty} \right] = 0 ,$$

$$\tag{II.8.26}$$

where $\varepsilon$ is infinitesimally small. The second part of the integral is

$$\int_0^\infty \frac{d\omega'}{\omega'^2 - \omega^2} \frac{\omega'\tau^*}{\sqrt{(\omega'\tau^*)^2 + 1}} . \tag{II.8.27}$$

We can evaluate this integral by introducing a new variable, $\xi^2 = (\omega'\tau^*)^2 + 1$, $\xi d\xi = \omega'\tau^{*2} d\omega'$. This integral then becomes

$$\frac{1}{\tau^*} \int_0^\infty \frac{d\xi}{\xi^2 - \tau^{*2}\omega^2 - 1} \tag{II.8.28}$$

$$= \frac{1}{\tau^*} \frac{1}{2\sqrt{\tau^{*2}\omega^2 + 1}} \ln\left( \frac{1 + \sqrt{\tau^{*2}\omega^2 + 1}}{-1 + \sqrt{\tau^{*2}\omega^2 + 1}} \right) . \tag{II.8.29}$$

When we multiply this result by the constant removed earlier,

$$\frac{2\omega}{\pi} \frac{\sqrt{2}e^2 l}{\pi \hbar a} , \tag{II.8.30}$$

we get back $\sigma_2(\omega)$ as given in Eq. II.8.17. We have proved that the conductivity, $\sigma(\omega)$, is indeed an analytical complex function.

## 8.6 Solution: Frequency-Dependent Response of a Superconductor

From the oscillator sum rule, we know that

$$\int_0^\infty \sigma_1(\omega) \, d\omega = \frac{ne^2}{m} \frac{\pi}{2} . \tag{II.8.31}$$

Using $\sigma_1(\omega)$ for the superconductor, as given for this problem, we obtain

$$\int_0^\infty A\delta(\omega) \, d\omega = \frac{1}{2}A \int_{-\infty}^\infty \delta(\omega) \, d\omega = \frac{1}{2}A . \tag{II.8.32}$$

Therefore, putting Eqs. II.8.31 and II.8.32 together, we solve for $A$:

$$A = \frac{ne^2\pi}{m} . \tag{II.8.33}$$

We use the Kramers–Kronig relation to obtain $\sigma_2(\omega)$ from $\sigma_1(\omega)$:

$$\sigma_2(\omega) = -\frac{2\omega}{\pi} \int_0^\infty \frac{\sigma_1(\omega')}{\omega'^2 - \omega^2} \, d\omega' \tag{II.8.34}$$

$$= \frac{A}{\pi\omega} = \frac{ne^2}{m\omega} . \tag{II.8.35}$$

We could have obtained the identical result by using a Drude model conductivity with $1/\tau \to 0$.

The reflectivity coefficient is defined as

$$R \equiv \left| \frac{\sqrt{\varepsilon} - 1}{\sqrt{\varepsilon} + 1} \right| , \tag{II.8.36}$$

and for the superconductor in this problem we obtain

$$\varepsilon = 1 - \frac{4A}{\omega^2} + iA\delta(\omega) . \tag{II.8.37}$$

At finite frequencies (more specifically $\omega < \omega_p$), $\frac{4A}{\omega^2} \sim 1$. Then $\mathrm{Re}(\sqrt{\varepsilon}) = 0$ and $\mathrm{Im}(\sqrt{\varepsilon}) \neq 0$. So the reflectivity coefficient will be $R = 1$. This means that at low frequencies the superconductor is a perfect reflector.

The penetration depth for $\omega \ll \omega_p$ is defined as (see Ziman [3] pp. 398–404)

$$\lambda \equiv \frac{|\delta|^2}{\mathrm{Im}(\delta)}, \quad \text{where} \quad \delta = \frac{c}{\omega\sqrt{\varepsilon}} . \tag{II.8.38}$$

For this problem, the penetration depth is

$$\lambda = \sqrt{\frac{mc^2}{\omega\pi m e^2}} , \tag{II.8.39}$$

which is equivalent to the London penetration depth.

## 8.7 Solution: Transmission of a Thin Superconductor

(a) According to the Kramers–Kronig relation for the real and imaginary part of the conductivity, we have

$$\sigma_2(\omega) = -\frac{2\omega}{\pi} \int_0^\infty \frac{\sigma_1(\omega')}{\omega'^2 - \omega^2}\, d\omega' \; . \tag{II.8.40}$$

Using $\sigma_1$ of a superconductor given in the problem, $\sigma_1 = A\delta(\omega)$, we obtain the imaginary part of the conductivity: $\sigma_2 = \frac{2}{\pi}\frac{A}{\omega}$.

Substituting these conductivities into the transmission equation,

$$t(\omega) = \frac{1}{(1 + \frac{dz}{2}\sigma_1)^2 + (\frac{dz}{2}\sigma_2)^2} = \frac{1}{1 + (\frac{dz}{2}\frac{2}{\pi}\frac{A}{\omega})^2} \; . \tag{II.8.41}$$

For small $\omega$, the transmission becomes

$$t(\omega) \approx \left(\frac{\pi}{dz}\right)^2 \frac{1}{A^2}\omega^2 \; . \tag{II.8.42}$$

Therefore if we plot $t$ versus $\omega^2$, we should get a straight line with a slope proportional to $1/\Omega_p^4$.

(b) From the sum rule we know that

$$\int_0^\infty \sigma_1(\omega')\, d\omega' = \frac{\pi}{2}\frac{ne^2}{m} = A \; . \tag{II.8.43}$$

We can solve this resulting in $n = \frac{2}{\pi}\frac{m}{e^2}A$, and we also know that $A = \frac{1}{\sqrt{t}}\frac{\pi}{dz}\omega$ from part (a).

Using CGS units, we obtain $z = \frac{4\pi}{c}$, and we solve for $n$ ($\omega = \frac{2\pi c}{\lambda}$):

$$n = \frac{2}{\pi}\frac{m}{e^2}\frac{1}{\sqrt{t}}\frac{\pi}{d}\frac{c}{4\pi}\frac{2\pi c}{\lambda} = \frac{mc^2}{e^2 d\lambda} \; . \tag{II.8.44}$$

Plugging in the given values $t = 0.02$, $\lambda = 10^{-2}$ cm, and $d = 10^{-5}$ cm and using $\frac{e^2}{\hbar c} = \frac{1}{137}$, we end up with

$$n = \frac{137}{d\lambda}\frac{mc}{\hbar} = 1.2 \times 10^{24} \text{ cm}^{-3} \; . \tag{II.8.45}$$

## 8.8 Solution: Bloch Oscillations

(a) According to the quasi-classical equation of motion, the electron wavenumber varies in time as $\hbar \dot{k} = eE$, where $\dot{k}$ is the first derivative of $k$. Therefore a constant field results in

$$k = eEt/\hbar,\qquad\qquad (\text{II}.8.46)$$

where $t$ is the time. The time required for any given electron to move from $-\pi/a$ to $\pi/a$ in the Brillouin zone is $T_0 = 2\pi\hbar/eEa$; this is the period of oscillation.

For $a = 50$ Å and $E = 5 \times 10^4$ V/cm the period is $T_0 = 1.6 \times 10^{-13}$ sec. This time is in the range of the typical relaxation time of the electrons. These oscillations could be observed experimentally, if materials of sufficient purity can be prepared in which the relaxation time is longer than the period of the oscillations. Note that the oscillation period varies inversely with the lattice spacing and the electric field. In simple metals the typical lattice spacing is shorter and $T_0$ is longer. The high electric field used here cannot be maintained in metals. Therefore Bloch oscillations are not expected to be seen in simple metals.

(b) The contribution to the polarization by one electron is $P = ex$, where $x$ is the position of the electron wavepacket. The group velocity of the electrons is $v = \frac{1}{\hbar}\frac{\partial E}{\partial k} = \hbar k/m$. As the wavenumber varies in time according to Eq. II.8.46, the electron velocity follows $v = eEt/m$ and the position of the wavepacket is $x = \frac{eE}{2m}t^2 \equiv x_t^2$. When $k$ reaches the Brillouin zone boundary at $\pi/a$, it jumps back to $-\pi/a$. Accordingly, the velocity $v(t)$ jumps from $v = v_0 \equiv \pi\hbar/(ma)$ to a negative value, $v = -v_0$. The velocity and displacement is illustrated in Figure II.8.4. Note that the maximum polarization, $P_0 = ex_0 =$

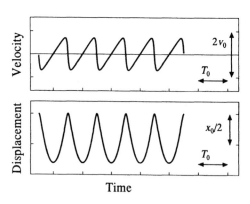

Fig. II.8.4. Velocity and displacement of a single electron moving in a weak periodic potential, under the influence of a constant electric field. The period of the motion is $T_0 = 2\pi\hbar/eEa$, the maximum velocity is $v_0 = \pi\hbar/ma$, and the displacement scale is $x_0 = h^2/(2meEa^2)$.

$h^2/2mEa^2$ is *inversely* proportional to the electric field and is independent of the electron charge.

To calculate the polarization of the electron system, one has to add up the contributions of each electron. For a large system of fixed electron density $n$,

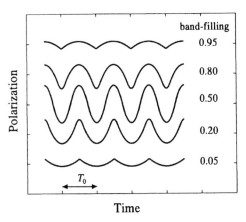

**Fig. II.8.5.** Polarization oscillations for various electron densities. The band-filling parameter is $na/2$, where the factor $1/2$ is due to the spin. The polarization scale is set by $P_0 = ex_0 = h^2/2mEa^2$.

the polarization will be proportional to the number of electrons, $N_e = Nn$, where $N$ is the number of sites. The quantity of importance is the polarization per site, defined as $P = \frac{1}{N}\sum ex_i$, where $e$ is the electron charge and $x_i$ is the position of the $i$th electron, similar to that indicated in Figure II.8.4. Adding up the contribution of the electrons is complicated by the fact that the electrons do not move "in phase": The electron on the right-hand side of the Fermi sea will reach the zone boundary first, and other electrons follow later. For very low electron densities the polarization will be similar to the displacement seen in the lower panel of Figure II.8.4, but as the band filling is increased the sharp features will be "rounded" (see Figure II.8.5). At the same time, the magnitude of the signal increases, since there are more electrons contributing. A large amplitude is reached if $n = 1/a$ – that is, for a half-filled band. When the band filling (calculated as $na/2$) approaches 1, the signal starts to drop, since some fraction of the electrons will move in a direction opposite to the others at any given moment in time. The oscillation disappears as the band becomes completely full.

# 9 Interactions and Phase Transitions

## 9.1 Solution: Spontaneous Polarization

In Problem 2.15 we investigated polarization waves in a similar system. That solution indicated that the instability will develop at $k = 0$. The equilibrium condition is given by

$$\frac{\partial^2 p}{\partial t^2} = 0 = -\omega_0^2 p_0 - \lambda p_0^3 + \frac{4\alpha\omega_0^2}{a^3} \sum_{m=1}^{N/2} \frac{1}{m^3} p_0 \,. \tag{II.9.1}$$

The sum converges fast to a value of 1.202. We now introduce the new parameter $\alpha_0 = a^3/4.808\omega_0^2$. $p_0 = 0$ is always a solution to this equation. The other possible solution is

$$p_0^2 = \frac{\omega_0^2}{\lambda}\left(\frac{\alpha}{\alpha_0} - 1\right)$$

$$p_0 = \pm\omega_0\sqrt{\frac{1}{\lambda}\left(\frac{\alpha}{\alpha_0} - 1\right)} \,. \tag{II.9.2}$$

For $\alpha < \alpha_0$, $p_0 = 0$ is the only solution. For $\alpha > \alpha_0$, the $p_0 = 0$ solution becomes unstable, and $p_0$ is given by Eq. II.9.2. A critical value of $\alpha$ separates the two regimes: $\alpha_{\mathrm{crit}} = \alpha_0 = a^3/4.808\omega_0^2$. Since $p_0 \sim (\alpha - \alpha_{\mathrm{crit}})^\beta$, Eq. II.9.2 gives $\beta = \frac{1}{2}$.

In real materials the electric polarizability (or magnetic susceptibility) of the constituents may be temperature-dependent in a simple manner: $\alpha \sim 1/T$. Inserting this into Eq. II.9.2 leads to a nonzero static polarization at low temperatures and zero average polarization at high temperatures. The critical temperature of the transition corresponds to the critical value of $\alpha$.

In this context our solution corresponds to the mean field approximation of the thermodynamic problem, since the fluctuations around the average value are neglected. In fact the exact solution would lead to zero average polarization at any nonzero temperatures, since in a one-dimensional system with short ranged interactions the fluctuation effects are always very strong.

## 9.2 Solution: Divergent Susceptibility

(a) The local field at position $x$ is the sum of the external field and the field created by the other molecules:

$$E = e^{ikx}e^{-i\omega t}\left(E^* + \frac{4p^*}{a^3}C(k)\right) , \qquad (II.9.3)$$

where $E^*$ is the amplitude of the externally aplied electric field, and $C(k) = \sum_{m=1}^{N/2}\frac{1}{m^3}\cos kam$, as seen in Eq. II.2.86.

Inserting this into the equation of motion (Eq. I.9.13), we obtain

$$-\omega^2 p^* = -\omega_0^2 p^* + \alpha\omega_0^2\left(E^* + \frac{4p^*}{a^3}C(k)\right) . \qquad (II.9.4)$$

Solving this equation to get the susceptibility, $\chi = p^*/E^*$,

$$\chi(k,\omega) = \frac{\alpha}{1 - \left(\frac{\omega}{\omega_0}\right)^2 - \frac{4\alpha}{a^3}C(k)} . \qquad (II.9.5)$$

At $\omega = 0$, this becomes

$$\chi(k) = \frac{\alpha}{1 - \frac{4\alpha}{a^3}C(k)} ; \qquad (II.9.6)$$

at $k = 0$, the susceptibility is

$$\chi(\omega) = \frac{\alpha}{1 - \frac{\alpha}{\alpha_{\text{crit}}}} , \qquad (II.9.7)$$

where $\alpha_{\text{crit}} = a^3/4.808\omega_0^2$ (see Solution 9.1). For $\alpha$ close to $\alpha_{\text{crit}}$ the expression can be expanded as $\chi \sim (\alpha_{\text{crit}} - \alpha)^\gamma$ with $\gamma = -1$. The $\chi(\alpha)$ function can be turned into a temperature-dependent susceptibility as discussed in Solution 9.1. Around the critical temperature the characteristic exponent will remain $\gamma = -1$.

(b) At $\alpha = \alpha_{\text{crit}}$ the $k$-dependence of the susceptibility is

$$\chi(k) = \frac{1}{1 - \frac{C(k)}{1.202}} . \qquad (II.9.8)$$

We now have to take a closer look at the function $C(k)$ given in Eq. II.2.86. The small-$k$ expansion is somewhat odd,

$$C(k) = \sum_m \frac{1}{m^3} - \frac{1}{2}\sum_m \frac{1}{m^3}(kam)^2 . \qquad (II.9.9)$$

The first term is just 1.202, but the second term becomes

$$\sum_{m=1}^{N} \frac{1}{m} = 0.577 + \ln N + \cdots , \qquad (\text{II}.9.10)$$

and $\ln N \to \infty$ for $N \to \infty$! However, for any fixed $k$, the expansion breaks down at large enough $m$, and the divergence is eliminated. To obtain an estimate we introduce an effective upper limit to the summation, determined by $kaN = \pi$. Equation II.9.9 then becomes

$$C(k) \approx 1.202 - \frac{1}{2}\left(0.577 + \ln|\frac{\pi}{ka}|\right)(ka)^2 . \qquad (\text{II}.9.11)$$

The susceptibility at $\alpha_{\text{crit}}$ is:

$$\chi(k) = \frac{1}{1 - \frac{C(k)}{1.202}} \approx \frac{2.404}{\left(0.577 + \ln|\frac{\pi}{ka}|\right)(ka)^2} . \qquad (\text{II}.9.12)$$

For small-$k$ the susceptibility diverges as $-1/(k^2 \ln|k|)$. Figure II.9.1 illustrates this behavior. The divergence of the susceptibility at $k = 0$ indicates

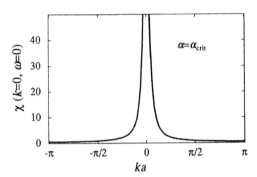

Fig. II.9.1. The susceptibility as a function of wavenumber at the critical value of $\alpha$. This plot is based on the exact $C(k)$ function. The approximate formula, Eq. II.9.12, results in a nearly identical graph.

that the instability of the system develops at this wavenumber. All electric dipole moments point in the same direction and the behavior is similar to ferromagnetism. This is not always true. For example, if the polarizations in this problem are aligned perpendicular to the chain, the divergence will occur at $k = \pi/a$ and an "antiferromagnetic" order develops.

## 9.3 Solution: Large-U Hubbard Model

In the Hubbard model, the Heisenberg model, and many other models used in solid state physics, a large class of wavefunctions belonging to higher energies are routinely neglected. Therefore the single-electron states are extremely simple: spin ↑ or spin ↓ in the Heisenberg model, or an electron with spin $\sigma$ at site $i$ in the Hubbard model. Compare that to the infinite number of energy levels in a hydrogen atom! This simplification is justified (since the thermal energies are usually much smaller than the atomic energy spacings) and necessary (otherwise we could not even begin to solve the many-body problem).

We want to solve the Schrödinger equations, drop the high-energy excitations, and match the low-energy part. In finite systems, like the two-electron problem we have here, the best way to find the energy spectrum is by diagonalizing the Hamiltonian matrix. Due to the simplicity of the one-electron problem, this matrix has a relatively small size and a simple structure.

The energy spectrum of the spin system is quite straightforward to determine. Let us choose first the basic vectors as $|\uparrow_1, \uparrow_2\rangle$, $|\downarrow_1, \downarrow_2\rangle$, $|\uparrow_1, \downarrow_2\rangle$, and $|\downarrow_1, \uparrow_2\rangle$. The subscripts indicate the site index, and the arrows are the spin states. Using the properties of spin operators (see Eq. I.5.14) and the fact that the $S_i$ acts only on spin $i$, we obtain the Hamiltonian matrix

$$H/J = \begin{pmatrix} 1/4 & 0 & 0 & 0 \\ 0 & 1/4 & 0 & 0 \\ 0 & 0 & 1/4 & -1/2 \\ 0 & 0 & -1/2 & 1/4 \end{pmatrix}. \tag{II.9.13}$$

The eigenvectors are

$$\begin{pmatrix} |\uparrow_1, \uparrow_2\rangle \\ |\downarrow_1, \downarrow_2\rangle \\ \frac{1}{\sqrt{2}}(|\uparrow_1, \downarrow_2\rangle + |\downarrow_1, \uparrow_2\rangle) \\ \frac{1}{\sqrt{2}}(|\uparrow_1, \downarrow_2\rangle - |\downarrow_1, \uparrow_2\rangle) \end{pmatrix}, \tag{II.9.14}$$

with energies $\frac{1}{4}J$, $\frac{1}{4}J$, $\frac{1}{4}J$, and $-\frac{3}{4}J$, respectively. Since the the total spin operator, $S = S_1 + S_2$, commutes with $H$, we may choose common eigenfunctions for $H$ and $S$; the states given above happen to belong to the spin states of $S = 1, S_z = 1$; $S = 1, S_z = -1$; $S = 1, S_z = 0$; and $S = 0, S_z = 0$; respectively. The energy difference between the higher energy $S = 1$ triplet states and the lower energy $S = 0$ singlet state is $J$. For the purposes of comparing the results with the Hubbard calculations, the absolute energy values are not necessary.

Let us now calculate the energy spectrum of the two-electron system. We have two sites (labeled as 1 and 2) and two spin states ($\uparrow$ and $\downarrow$). The electron Hamiltonian is

$$\begin{aligned} H &= t(c_{1,\uparrow}^+ c_{2,\uparrow} + c_{2,\uparrow}^+ c_{1,\uparrow} + c_{1,\downarrow}^+ c_{2,\downarrow} + c_{2,\downarrow}^+ c_{1,\downarrow}) \\ &\quad + U(c_{1\uparrow}^+ c_{1\uparrow} c_{1\downarrow}^+ c_{1\downarrow} + c_{2\uparrow}^+ c_{2\uparrow} c_{2\downarrow}^+ c_{2\downarrow}) \end{aligned} \tag{II.9.15}$$

We can express the six possible electronic states by acting on the vacuum state $|0\rangle$ with the electron creation operators:

$$|1\rangle = c_{1\uparrow}^+ c_{1\downarrow}^+ |0\rangle$$
$$|2\rangle = c_{2\uparrow}^+ c_{2\downarrow}^+ |0\rangle$$
$$|3\rangle = c_{1\uparrow}^+ c_{2\downarrow}^+ |0\rangle \qquad \text{(II.9.16)}$$
$$|4\rangle = c_{2\uparrow}^+ c_{1\downarrow}^+ |0\rangle$$
$$|5\rangle = c_{1\uparrow}^+ c_{2\uparrow}^+ |0\rangle$$
$$|6\rangle = c_{1\downarrow}^+ c_{2\downarrow}^+ |0\rangle \ .$$

Note that in a more realistic "two-atom" system we should also consider the $|7\rangle = c_{1\uparrow}^+ c_{1\uparrow}^+ |0\rangle$ and $|8\rangle = c_{2\uparrow}^+ c_{2\uparrow}^+ |0\rangle$ states, involving electrons on the same site. In the Hubbard model we do not allow this possibility, since the second electron can only occupy a higher atomic energy level.

Let us calculate the Hamiltonian matrix elements. For example, let us look at

$$\langle 1|H|1\rangle = \langle 0|c_{1\uparrow} c_{1\downarrow}|H|c_{1\downarrow}^+ c_{1\uparrow}^+|0\rangle \ . \qquad \text{(II.9.17)}$$

The standard way of evaluating this type of equation is to rearrange the electron operators such that the rightmost one is an annihilation operator, thereby giving zero contribution. In the process we have to use the commutation relation $c_{i,\sigma}^+ c_{j,\sigma'} + c_{j,\sigma'} c_{i,\sigma}^+ = \delta_{i,j}\delta_{\sigma,\sigma'}$. When creation and an annihilation operators of the same site and spin are interchanged, we are left with a nonzero $c$ term. The procedure is rather awkward, but it becomes easy to discover a systematic behavior. Finally we obtain

$$H = \begin{pmatrix} U & 0 & t & t & 0 & 0 \\ 0 & U & t & t & 0 & 0 \\ t & t & 0 & 0 & 0 & 0 \\ t & t & 0 & 0 & 0 & 0 \\ 0 & 0 & 0 & 0 & 0 & 0 \\ 0 & 0 & 0 & 0 & 0 & 0 \end{pmatrix} \ . \qquad \text{(II.9.18)}$$

Solving the Schrödinger equation means finding the diagonal representation of this matrix. Even for as little as two electrons, we have a $6 \times 6$ matrix! Diagonalizing a $6 \times 6$ matrix is no simple undertaking, but this one has many zeros and symmetries.

The structure of the matrix clearly indicates that the first four and the last two states split to two separate groups. For the last two states the eigenvectors are $|5\rangle$ and $|6\rangle$, with degenerate energies $E = 0$. As we start working on the remaining $4 \times 4$ matrix, we recognize that introducing two new basis vectors $|3'\rangle = \frac{1}{\sqrt{2}}(|3\rangle + |4\rangle)$ and $|4'\rangle = \frac{1}{\sqrt{2}}(|3\rangle - |4\rangle)$ further simplifies the problem, leading to

$$H = \begin{pmatrix} U & 0 & 2t & 0 & 0 & 0 \\ 0 & U & 0 & 0 & 0 & 0 \\ 2t & 0 & 0 & 0 & 0 & 0 \\ 0 & 0 & 0 & 0 & 0 & 0 \\ 0 & 0 & 0 & 0 & 0 & 0 \\ 0 & 0 & 0 & 0 & 0 & 0 \end{pmatrix} \ . \qquad \text{(II.9.19)}$$

The last three states now all have $E = 0$. The first three states have a more complicated spectrum, obtained from the determinant

$$\det \begin{vmatrix} U - E & 0 & 2t \\ 0 & U - E & 0 \\ 2t & 0 & -E \end{vmatrix} = 0 . \tag{II.9.20}$$

The results are $E_{\pm} = U/2 \pm \sqrt{4t^2 + U^2/4}$. For large $U$ the two values are $E_+ \approx U$ and $E_- \approx -4t^2/U$. The eigenvector for the low-energy state is $|\text{low}\rangle = \frac{1}{1+(E_-/2t)^2}|3'\rangle + \frac{E_-}{2t}\frac{1}{\sqrt{2}}(|1\rangle + |2\rangle) \approx \frac{1}{1+4t^2/U^2}|3'\rangle + 2t/U|1'\rangle$, where $|1'\rangle = \frac{1}{\sqrt{2}}(|1\rangle+|2\rangle)$. The low-energy state is mostly $|3\rangle$ and $|4\rangle$ with a little bit of $|1\rangle$ and $|2\rangle$ mixed in. Remember, in $|3\rangle$ and $|4\rangle$ the electrons are on different sites, and in $|1\rangle$ and $|2\rangle$ they are on the same site. The energy is lowered if the electrons are not restricted entirely to two separate sites – similar to the kinetic energy gain in every case when the electron is delocalized.

Let us now look at all lower-energy states. Altogether, we have four states to consider: $|\text{low}\rangle$ with energy $-4t/U^2$, and $|4'\rangle$, $|5\rangle$, and $|6\rangle$ with energy zero. Compare this to the solution of the spin Hamiltonian. If we choose $J = 4t^2/U$ we can match the energy splitting, the degeneracies, and even the total spin states.

Extending the Hubbard model calculation to a more than two sites dramatically increases the complexity of the problem. Nevertheless, computer calculations on finite systems are often done with a philosophy similar to this solution. It is worthwhile to note that exact solutions for the $d$-dimensional Hubbard model only exist for $d = 1$ and $d \to \infty$.

A more general, related problem (the "$t-J$" model) is discussed by Harris and Lange.[1]

## 9.4 Solution: Infinite Range Hubbard Model

Using the commutation relations (see Eq. I.5.3), it is easy to show that $n_{\uparrow}$ and $n_{\downarrow}$ commute with H. Therefore any solution of the Schrödinger equation, which satisfies $H|\psi\rangle = E|\psi\rangle$, must also satisfy $n_{\uparrow}|\psi\rangle = n_{\uparrow}|\psi\rangle$ and $n_{\downarrow}|\psi\rangle = n_{\downarrow}|\psi\rangle$, where $n_{\uparrow}$ and $n_{\downarrow}$ are numbers. In other words, each energy can be labeled according to the number of "up" and "down" spin states in the system, $n_{\uparrow}$ and $n_{\downarrow}$.

The operator $a^+_{\vec{k},\uparrow}$ will create the solutions of the Schrödinger equation if $[H, a_{\vec{k},\uparrow}] = E_{\uparrow}(\vec{k})$ is satisfied, where $E_{\uparrow}(\vec{k})$ is the energy of the corresponding solution. After evaluating another round of elementary commutators, one obtains $E_{\uparrow}(\vec{k}) = E(\vec{k}) + Un_{\downarrow}$ and $E_{\downarrow}(\vec{k}) = E(\vec{k}) + Un_{\uparrow}$.

Note that the mean field approximation and the exact solution are equivalent for this problem. This feature demonstrates an interesting paradox in

---

[1] A.B. Harris and R.V. Lange, *Phys. Rev.*, **157**, 295 (1967).

the theory of interacting many-body systems: While there is no general solution for an arbitrary interaction potential (just try it!), an exact solution is possible for the infinite range interaction (or for no interaction at all).

## 9.5 Hint: Stoner Model

The Stoner model is discussed by Ziman [3] pp. 339–340. The susceptibility is $\chi = \mu^2 g(E_F)/(1 - U g(E_F)/2)$. For $U = 0$ the result reproduces the free-electron expression. For $U > 0$ the susceptibility is enhanced (this is the *Stoner enhancement*), and at $U = 2/g(E_F)$ it diverges. At this point the system develops a spontaneous magnetization, and it is ferromagnetic for larger values of $U$.

## 9.6 Solution: One-Dimensional Electron System

In one dimension the dielectric function is

$$\epsilon = 1 - U(q)\frac{1}{L}\sum_{k,\sigma} \frac{f(k+q) - f(k)}{E(k+q) - E(k)} , \qquad (\text{II}.9.21)$$

where $q_x \equiv q$, $L$ is the length of the system, $k$ is the wavenumber, and $\sigma$ is the spin. For free-electrons the energy is given by $E = \hbar^2 k^2/2m$. We replace the sum in Eq. II.9.21 by an integral:

$$\sum_{k\sigma} \Rightarrow 2 \int \frac{dk}{2\pi/L} . \qquad (\text{II}.9.22)$$

For a system like this, where the the properties are independent of spin, the

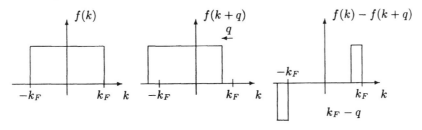

Fig. II.9.2 Fermi functions of a one-dimensional system for the calculation of $\epsilon$.

sum over the spin states results in a factor of two. At zero temperature the Fermi function is a step function. Figure II.9.2 illustrates how Fermi functions will influence the dielectric function calculation.

Eq. II.9.21 becomes

$$\epsilon = 1 - U(q) \frac{1}{\pi} \left[ \int_{-k_F-q}^{-k_F} \frac{dk}{\frac{\hbar^2}{2m}(2kq + q^2)} \right.$$

$$\left. - \int_{k_F-q}^{k_F} \frac{dk}{\frac{\hbar^2}{2m}(2kq + q^2)} \right] .$$

(II.9.23)

The integration then yields

$$\epsilon = 1 + \frac{U(q)}{\pi} \frac{2m}{\hbar^2} \frac{1}{2q} \ln \left( \frac{2k_F + q}{2k_F - q} \right)^2 .$$

(II.9.24)

Therefore we obtain

$$\epsilon = 1 - U(q)\chi(q)$$

(II.9.25)

with

$$\chi = -\frac{1}{\pi q} \frac{2m}{\hbar^2} \ln \left| \frac{2k_F + q}{2k_F - q} \right| = -\frac{g(E_F)}{q} \ln \left| \frac{2k_F + q}{2k_F - q} \right| ,$$

(II.9.26)

where the density of states at the Fermi level, $g(E_F)$, was introduced. The Linhard function $\chi(q)/g(E_F)$ is plotted in Figure II.9.3.

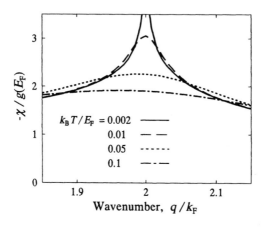

**Fig. II.9.3.** The wavenumber dependence of the Linhard function in one dimension for various temperatures.

Around $q \approx 2k_F$, we get $\epsilon \approx 1 + Ug[\ln|4k_F/(2k_F - q)|]$. For $U < 0$, the logarithmic divergence of the term in square brackets ensures that $\epsilon = 0$ is reached at arbitrarily weak interactions. The negative interaction energy means an attractive interaction. Phonons can mediate attractive interactions between electrons in a metal. The ground state of this system is *not* homogeneous in the charge density, but in fact it has a $q = 2k_F$ charge density wave (CDW). The strong divergence of $\chi$, and the tendency for instability, is a property of the one-dimensional electron system (see Problem 9.8).

At finite temperatures the logarithmic divergence is removed, and the metallic state becomes stable. This is illustrated in Figure II.9.3 made from the numerical integration of Eq. II.9.21 at finite temperatures. The temperature dependence is discussed in detail in Problem 9.9.

One may wonder what happens for repulsive interactions, $U(q) > 0$? The dielectric function defined in Eq. II.9.21 will have a positive divergence, instead of reaching zero. There will be no CDW instability; in fact the external potential will be very effectively shielded. However, the stability in the CDW "channel" does not mean that the system does not have other types of instabilities. In particular, superconductivity and spin density waves may occur. A careful treatment of this problem requires going beyond second-order perturbation theory; for further details, see the review by Sólyom.[2]

## 9.7 Solution: Peierls Distortion

(a) For $k_B T \ll \Delta$, the calculation has already been performed in the solution to Problem 6.8. The result is

$$\delta E \approx \frac{1}{2}\frac{\Delta^2}{W} \log \Delta/W < 0 . \tag{II.9.27}$$

As the gap opens, the energy change is always negative since the argument within $\log \Delta/W$ is less than one. The entropy contribution was shown to be negligible in Solution 6.8.

For $k_B T \gg \Delta$, Eq. II.9.28 can be used. The change of the free energy is

$$\delta F \equiv F(\Delta) - F_0 = \delta E - T\delta S = L \int (g - g_0) f(\mathcal{E} - 2k_B T \log f) d\mathcal{E} , \tag{II.9.28}$$

where $F_0 \equiv F(\Delta = 0)$. The appropriate high temperature expansion of the Fermi function yields $f \approx \frac{1}{2}(1 - \frac{\mathcal{E}}{2k_B T})$ and $\log f = -\log 2 - \frac{\mathcal{E}}{2k_B T}$. The even terms in the integrand will have no contribution. The constant term $(2k_B T \log 2)$ does not contribute either, since $\int (g - g_0) d\mathcal{E} = 0$. At high temperatures the first non-vanishing term in the free energy is

$$\delta F \approx g_0 W \frac{1}{4}\frac{\Delta^2}{k_B T} \sim \frac{\Delta^2}{k_B T} > 0 , \tag{II.9.29}$$

In summary, a gap much larger than $k_B T$ is favorable, as long as the other conditions in this problem are satisfied (in particular, $\Delta$ must be much less than $W$). The opening of a small gap increases the free energy.

(b) The change in the total energy is given by

$$\delta E_{tot} = \approx \frac{1}{2}\frac{\Delta^2}{W} \log \Delta/W + \frac{1}{2}\kappa s^2 = \frac{1}{2}\Delta^2 \left( \frac{1}{W} \log \Delta/W + \kappa/\alpha \right) ;. \tag{II.9.30}$$

---

[2] J. Sólyom, *Adv. Phys.* **28**, 201 (1979).

For small, but finite $\Delta$ the term in parenthesis is always negative: The energy gain in the electron system is always larger the energy investment to the lattice distortion. Therefore the development of the lattice distortion will happen spontaneously, a finite gap will develop and the metallic state is unstable. This is the mechanism of the *Peierls transition*, which has been experimentally observed in quasi-one-dimensional solids.

The magnitude of the gap is determined from the equilibrium condition

$$\frac{\delta F}{\delta \Delta} = 0 = \Delta \left( \frac{1}{W} \log \Delta/W + \kappa/\alpha \right) + \frac{1}{2}\Delta . \tag{II.9.31}$$

The result is $\Delta = We^{-1/\lambda}$, where $\lambda = \alpha/W\kappa$ is the dimensionless electron–phonon coupling constant. Note that the result is singular around $\lambda = 0$; all coefficients of the Taylor series are zero.

Mathematically similar results are encountered in the BCS theory of superconductivity.[3]

(c) We leave the solution of this part to the reader.

## 9.8 Hint: Singularity at $2k_F$

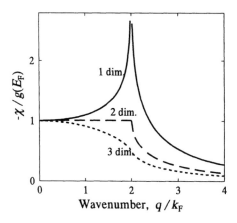

**Fig. II.9.4.** The wavenumber dependence of the Linhard function in 1, 2, and 3 dimensions and at zero temperature.

The results are:

$$\begin{aligned} \text{1 dimension} \quad & -\chi/g(E_F) = \frac{1}{q}\ln\left|\frac{2k_F + q}{2k_F - q}\right| , \\ \text{2 dimensions} \quad & -\chi/g(E_F) = 1 \qquad \text{for } q < 2k_F \text{ and} \end{aligned} \tag{II.9.32}$$

---

[3] See Tinkham [20] pp.30–35.

$$= 1 - \frac{\sqrt{q^2 - (2k_F)^2}}{q} \quad \text{for } q > 2k_F \, ,$$

$$\text{3 dimensions} \quad -\chi/g(E_F) = \frac{1}{2} + \frac{(2k_F)^2 - q^2}{8qk_F} \ln \left| \frac{2k_F + q}{2k_F - q} \right| \, ,$$

where $q = |\mathbf{q}|$. Note the negative sign of the susceptibility in all dimensions. These functions are plotted in Figure II.9.4. The one-dimensional result was derived in more detail in Solution 9.6.

## 9.9 Solution: Susceptibility of a One-Dimensional Electron Gas

For plane wave electronic wavefunctions,[4] Eq. I.9.2 results in a susceptibility

$$\chi = 2 \int dk \frac{f(k+q) - f(k)}{E(k+q) - E(k)} = -2 \int dk' \frac{f(E^-) - f(E^+)}{\frac{\hbar^2}{2m} 4k' k_F} \, , \qquad (\text{II.9.33})$$

where the factor of 2 comes from the sum over spin states, $E^{\pm} = \frac{\hbar^2}{2m} k'^2 + E_F \pm 2\frac{\hbar^2}{2m} k' k_F$, and we introduced $k' = k + k_F$ and $q = 2k_F$. The singularity of this integral is due to the divergence of the denominator. Figure II.9.5 illustrates that the critical regime is close to $k' = 0$. At low temperatures the difference between the Fermi functions turns out to be very close to $f(E^-) - f(E^+) = \tanh \frac{\hbar^2}{2m} \frac{k' k_F}{k_B T}$ for $|k'| < 2k_F$, with a sharp cut-off for $|k'| > 2k_F$.

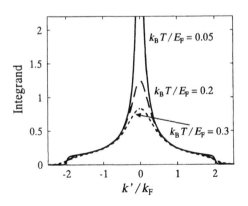

Fig. II.9.5. The wavenumber dependence of the integrand in Eq. II.9.33 for various temperatures

The integral to calculate is

$$\chi = \frac{2m}{\hbar^2 k_F} \frac{1}{2} \int_{-x_0}^{x_0} dx \frac{\tanh x}{x} \, , \qquad (\text{II.9.34})$$

---

[4] If the wavefunctions are not plane waves, the matrix element $\langle k + q | e^{ikx} | k \rangle$ will lead to an additional (smooth) dependence on $k$ and $q$.

where $x = \frac{\hbar^2}{2m} \frac{k' k_F}{k_B T}$ and $x_0 = 2E_F/k_B T$. For $x_0 \gg 1$ the integral can be performed as $\int_0^{x_0} dx/x \tanh x = \ln x \tanh x|_0^{x_0} - \int \ln x/\cosh^2 x\, dx \approx \ln x_0 + C - \ln(\pi/4)$. Therefore the low temperature susceptibility is

$$\chi = -g \ln \frac{1.13 E_F}{k_B T} \tag{II.9.35}$$

where $g = \frac{2m}{\hbar^2 k_F}$ is the density of states.

## 9.10 Solution: Critical Temperature in Mean Field Approximation

The susceptibility of the one-dimensional electron gas was determined in Problem 9.9. For $U < 0$ the $1 - U\chi$ condition results in a critical temperature of $T_c = (1.13 E_F/k_B) \exp(-1/|U|g)$. The electron system develops a charge density wave (CDW) at low temperatures.

It is important to emphasize that the solution is based on the mean field theory, and the entropy of the system was left out of the calculation. The exact solution (see Emery [24]) indicates that there is no phase transition in the one-dimensional system at any finite temperature and the CDW phase is only stable at $T = 0$. However, the fluctuations towards the CDW phase become quite strong at temperatures below the mean field $T_c$, and a snapshot of the system, taken at any given moment, looks very much like the CDW phase.

## 9.11 Solution: Instability of Half-Filled Band

Introducing integration to replace the sum over $k$-states we obtain

$$G(q) = -\int \frac{f(E(k)) - f(E(k+q))}{(E(k) - E(k+q))}\, d^2k . \tag{II.9.36}$$

With the energy dispersion specified in the problem, for a half-filled band the Fermi energy is zero. Also, at $q = q_0 \equiv (\pi/a, \pi/a)$ we have $E(k+q) = -E(k)$. Therefore

$$G(q = q_0) = \int \frac{f(E(k)) - f(-E(k))}{2E(k)}\, d^2k . \tag{II.9.37}$$

Since the integrand depends on $k$ through the energy only, it is practical to turn the integral over $k$ into an energy integral. Doing this brings the density of states $g(E)$ into the integrand:

$$G(q = q_0) = \int \frac{g(E)(f(E) - f(-E))}{2E}\, dE . \tag{II.9.38}$$

In order to investigate the characteristic temperature dependence, we replace the $f(E) - f(-E)$ function with $E/2k_BT$ for $-k_BT/2 < E < k_BT/2$ and $\pm 1$ otherwise (see Figure II.9.6). For energies larger than $k_BT$ the integrand becomes $g(E)/|E|$; for a $k_BT$ wide range around $E = 0$ it is simply $g(E)/2k_BT$.

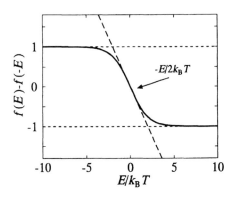

**Fig. II.9.6.** The approximation used to replace the Fermi function.

The density of states relevant here is investigated in Problem 4.3; for low energies $g(E)$ was found to be $\sim -\log(E/E_0)$. Consequently,

$$G(q = q_0) \sim \int_0^{k_BT} dE\, \frac{\log(E_0/E)}{2k_BT} + \int_{k_BT}^{E_0} dE \log(E_0/E)/E$$

$$= 1/2\log(E_0/k_BT) + 1/2\log^2(E_0/k_BT) . \qquad (II.9.39)$$

The leading divergence goes as $\log^2 T$. Note that $G(q)$ at $q = 0$ is also divergent as $Q(q = 0, T = 0) \sim g(E_F)$. It is easy to show that at this point the temperature dependence diverges as $\log T$.

## 9.12 Solution: Screening of an Impurity Charge

The potential, $\phi(q)$, set up in response to an applied potential, $\phi^{ext}(q)$, is expressed as follows (see Eq. I.9.3):

$$\phi(q) = \frac{1}{\epsilon(q)}\phi^{ext}(q) = \frac{1}{\epsilon(q)}\frac{4\pi Q}{q^2} , \qquad (II.9.40)$$

where $Q$ is the impurity charge. We used the Fourier transform of the Coulomb potential

$$\phi^{ext}(q) = \int d^3r\, e^{iq\cdot r}\frac{Q}{r} = \frac{4\pi Q}{q^2} . \qquad (II.9.41)$$

Let us introduce $\lambda^2 = Lk_0^2/q^2$. According to the approximation given in this problem, we obtain the dielectric function,

$$\epsilon(q) = \begin{cases} 1 + \frac{\lambda^2}{q^2} & \text{for } q < 2k_F \\ 1 & \text{for } q > 2k_F \ . \end{cases} \tag{II.9.42}$$

To calculate the total potential, $\phi(q)$, we use the Fourier transform,

$$\begin{aligned} \phi(r) &= \int \frac{4\pi Q}{q^2 \epsilon(q)} e^{iq \cdot r} \frac{d^3 q}{(2\pi)^3} \\ &= \frac{4\pi}{(2\pi)^3} Q \int_0^{2\pi} d\phi \int_{-1}^1 \int_0^{\infty} q^2 dq \ d\cos\theta \frac{1}{\epsilon(q)q^2} e^{iqr\cos\theta} \\ &= \frac{8\pi^2}{(2\pi)^3} \frac{Q}{r} \int_0^{\infty} \frac{\sin qr}{q\epsilon(q)} dq = \frac{1}{\pi} \frac{Q}{r} \int_0^{\infty} F(q) \sin qr \ dq \ . \end{aligned} \tag{II.9.43}$$

In the last step, we have defined the function,

$$F(q) = \begin{cases} \frac{q}{q^2 + \lambda^2} & \text{for } q < 2k_F \\ 1/q & \text{for } q > 2k_F \ . \end{cases} \tag{II.9.44}$$

$F(q)$ is sketched in Figure II.9.7. We can calculate the magnitude of the

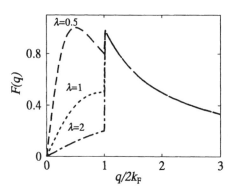

Fig. II.9.7. The function $F(q)$ for several values of $\lambda$.

jump in $F(q)$ at $q = 2k_F$ to be

$$\Delta F = \frac{2k_F}{(2k_F)^2 - \lambda^2} - \frac{1}{2k_F} = \frac{1}{2k_F} \frac{\left(\frac{\lambda}{2k_F}\right)^2}{1 - \left(\frac{\lambda}{2k_F}\right)^2} \ . \tag{II.9.45}$$

The asymptotic behavior of $\phi(r)$ is strongly influenced by the jump in $F(q)$. We can partially complete the integration of Eq. II.9.43:

$$\begin{aligned} &\int_0^{\infty} F(q) \sin qr \ dq \\ &= F(q)\frac{1}{r}(-\cos qr)\Big|_{q=0}^{\infty} - \int_0^{\infty} F'(q)\frac{-\cos qr}{r} dq \ . \end{aligned} \tag{II.9.46}$$

The first term evaluates to zero. $F'(q)$ contains $\delta(q - 2k_F)\Delta F$ as well as the smooth parts from $F(q)$. The integral of the smooth parts will average out with the oscillating $\cos qr$ function (as $r$ gets large, the oscillations will become very fast). The important part of the integral will come from $\delta(q - 2k_F)\Delta F$. We finally obtain

$$\phi(r) \approx \frac{1}{\pi} \frac{Q}{r} \frac{\cos 2k_F r}{r} \Delta F \sim \frac{\cos 2k_F r}{r^2} . \tag{II.9.47}$$

The asymptotic cutoff is faster is the singularity in $F(q)$ is weaker. The Lindhard function in three dimensions has a point of inflection at $2k_F$ and the corresponding Friedel oscillation decays as $\sim 1/r^3$.

## 9.13 Solution: Fermi Surface Nesting in Two Dimensions

Due to the energy denominator in the integrand, the peak positions of $G(q)$ can be determined by searching for a good match between the original Fermi surface and another one shifted by $q$. The Fermi surface (in two dimensions, the Fermi line) for a nearly half-filled system is shown in Figures II.3.11 or II.4.1. For simplicity let us first confine the $q$ to the first quadrant: $q_x \geq 0$, $q_y \geq 0$. Figure II.9.8a gives us a hint as to where to find good matches. For example, in the direction $q = (1,1)$, the best match happens if the length of the $q$ vector is just equal to the the "calipered" width of the Fermi surface, as demonstrated in Figure II.9.8b.

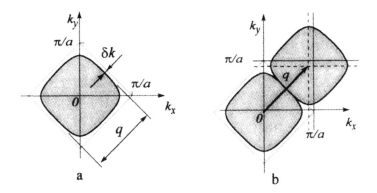

a                                b

**Fig. II.9.8.** (a) Typical "size" of the Fermi surface and (b) displacement of the electrons in the (1,1) direction by wavevector $q$ for optimum matching of the Fermi surfaces.

However, an even better match is found if we note that all considerations must be done in the extended zone scheme: We can wedge the displaced Fermi surface between the original Fermi surface and the one in the next

Brillouin zone (Figure II.9.9). Note that one of the components (let us say $q_y$) of $q$ becomes exactly $\pi/a$. We will now determine the other component in terms of the difference, $\delta q = \pi/a - q_x$. Let us first look at the separation

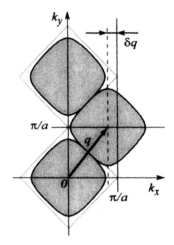

**Fig. II.9.9.** Matching of Fermi surfaces in the extended zone scheme.

of the Fermi line from the straight line corresponding to half-filled band (Figure II.9.8a). If we express $k_x$ and $k_y$ in terms of the length of $k$ vector in the (1,1) direction $k_x = k_y = \frac{1}{\sqrt{2}}\kappa$, we obtain $E(\kappa) = (-\cos(1/\sqrt{2})\kappa + 1)E_0/2$. Expanding in the neighborhood of $E_0/2$ we obtain $\delta k = \sqrt{2}2\delta E/E_0$, where $\delta E = E_0 - E_F$. "Wedging" occurs if the shift in the (1,0) direction is $\sqrt{2}$ times larger than $\delta k$. Finally, we should not forget that the displaced Fermi line has also shrunk; this brings in another factor of 2. We obtain $\delta q = 8\delta E/E_0$; $\delta q = 0.8$ with the numerical values given in the problem. Due to symmetry, peaks occur at the eight position seen in Figure I.9.3. Antiferromagnetic fluctuations in quasi-two-dimensional copper-oxide-based compounds like $La_{1-x}Sr_xCuO_4$ are believed to be due to the peaks found in the susceptibility.[5] Note that, according to the above result, the spin order does not correspond to alternating "up" and "down" spins on the lattice sites; the periodicity corresponding to similar peaks in the susceptibility is incommensurate with the crystal lattice.

---

[5] For experiments, see S.W. Cheong *et al.*, *Phys. Rev. Letters*, **67**, 1791 (1991); for theory, see N.Bulut, D. Hone, D.J. Scalapino, G. Bychkov, *Phys. Rev. Letters*, **64**, 2723 (1990).

## 9.14 Solution: Fermi Surface Nesting in Quasi One Dimension

We approximate the susceptibility by

$$\chi(q) \approx \frac{1}{V} \sum_{k,\sigma} \frac{f(k+q) - f(k)}{E(k+q) - E(k)} \equiv -G(q) , \qquad (\text{II.9.48})$$

where $V$ is the volume and $f(k) = f(E(k))$ is the Fermi function.[6] If the $q$-vector shifts the electronic states such that a good Fermi surface nesting is created, then the divergent energy denominator in Eq. II.9.48 will lead to peaks in $G(q)$.

Let us first look at $\alpha = 0$: The electron transfer in the $y$ direction vanishes so the system is one-dimensional. The energy surface is described by a simple cosine wave of amplitude $E_0$; the Fermi lines are at $k_x = \pm k_F$ with $k_F a = \arccos(E_F/E_0)$. The susceptibility is divergent at $q_0 = (2k_F, q_y)$ for any value of $q_y$, like in problem 9.6.

For small $\alpha$ the Fermi line is $k_x = \pm(k_F + \kappa(k_y))$, where $\kappa$ is small, with $k_F a = \arccos(E_F/E_0)$.[7] The Fermi lines look very much like weak cosine waves (see Figure I.9.4). Therefore a displacement by $q$ parallel to the the $x$ direction does not lead to good overlap between the Fermi lines, as illustrated in Figure II.9.10. Inspection of the figure reveals that a better match could be obtained by $q_0 = (\pm 2k_F, \pm \pi/b)$.

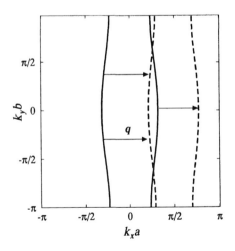

Fig. II.9.10. Displacement of the electronic states by wavevector $q$.

The $q_0 = (\pm 2k_F, \pm \pi/b)$ value would lead to a perfect match only if the Fermi lines were perfect cosine waves, $\kappa \propto \cos k_y b$. In reality, the Fermi line

---

[6] For a discussion of the validity of this approach, see problem 9.11.

[7] To obtain this value for the parameter $k_F$, one has to show that the area between the two Fermi lines (the total number of electrons) does not change much as $\alpha$ is increased from zero.

is more complicated. Let us expand the energy function in the neighborhood of $k_F$; $k_x = k_F + \kappa$. To second-order we obtain

$$E = -E_0(-\cos k_F a + a\kappa \sin k_F a + \frac{1}{2}a^2\kappa^2 \cos k_F a - \alpha \cos k_y b) . \quad (\text{II}.9.49)$$

Since the energy must be constant, $\kappa$ is obtained from

$$a\kappa \sin k_F a + \frac{1}{2}a^2\kappa^2 \cos k_F a - \alpha \cos k_y b = 0 . \quad (\text{II}.9.50)$$

A second-order expansion gives

$$\kappa = \frac{\alpha}{a \sin k_F a} \cos k_y b - \frac{\alpha^2}{2a} \frac{\cos k_F a}{\sin^3 k_F a} \cos^2 k_y b . \quad (\text{II}.9.51)$$

The two Fermi lines are described by $k_x^{(1)} = k_F + \kappa(k_y)$, and $k_x^{(2)} = -[k_F + \kappa(k_y)]$. We want to find a $q_0 = (q_{0x}, q_{0y})$ so that the shifted Fermi line, $k_x^{(2)} + q_{0x} = -[k_F + \kappa(k_y + q_{0y})]$ is as close as possible to $k_x^{(1)}$.

Two functions match perfectly if all of the coefficients in their Taylor series are equal. In this problem, we have two parameters to determine (the two components of $q_0$), and we have a free choice for the value of $k_y$, where the matching happens. The three numbers will be determined by requiring that the two functions, and their first- and second-derivatives, are equal. This condition leads to

$$q_{0x} = \pm\left[2k_F - \frac{1}{a}\frac{\alpha^2 \cos k_F a}{\sin^3 k_F a}\right] \quad (\text{II}.9.52)$$

$$q_{0y} = \pm\left[\frac{\pi}{b} - \frac{\alpha \cos k_F}{b \sin^2 k_F}\right] .$$

Note that the deviation of $q_{0x}$ from $2k_F$ is second-order in $\alpha$, while the correction in $q_{0y}$ is proportional to $\alpha$.

Throughout our solution, we assumed that the Fermi energy is fixed, but we varied the shape of the Fermi surface. Since the Fermi energy is determined by the number of electrons (the band filling), it would change as the Fermi surface is modified. This leads to a small correction to our result. The band filling is proportional to the area within the Fermi lines (in three dimensions, the volume of the Fermi surface). Therefore, as we consider the second-order corrections to the Fermi lines we have to change the Fermi energy in our calculation, leading to another term in $\delta q_x$ of $\sim \alpha^2$.

Figure II.9.11 shows the result of numerical integration of Eq. II.9.48 for a half-filled band and $\alpha = 0.1$. The function exhibits two "crests", running in the neighborhood of $\pm 2k_F$. Maxima indeed develop at the four $q$-points determined above.

The wavenumbers of the charge density waves in TaS$_3$ and the spin density waves in the organic compound BEDT TCNQ may be related to the peaks in the susceptibility determined here. Note that the corresponding wavelength is not a simple rational multiple of the lattice spacing, and the wave is *incommensurate* to the underlying lattice.

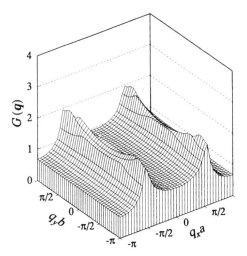

**Fig. II.9.11.** The electron susceptibility for a quasi-one-dimensional electron system, as defined by Eq. II.9.48. The band is half-filled, and $\alpha = 0.1$. The numerical integration was performed with a small, but finite temperature. As the temperature is lowered the function diverges at four values of $q$, calculated in the text.

## 9.15 Solution: Anderson Model

(a) If the hybridization term is zero, the Hamiltonian describes two independent systems: Band electrons and $d$ electrons. The behavior of the band electrons is trivial. The $d$ electron system splits into $N_d$ independent sites. The physical properties are determined by the relationship between $E_F$, $E_d$, and $E_d + U$. For $E_d < E_F$ and $E_d + U < E_F$ the $d$ sites are fully occupied; for $E_d > E_F$ and $E_d + U > E_F$ the $d$ sites are empty. There are no unpaired magnetic moments, and the system is diamagnetic.

For $E_d < E_F$ and $E_d + U > E_F$ the minimum energy configuration will have as many unpaired electrons on the $d$ sites as possible. This means that the system will have magnetic moments on most sites; when $N_d^e = N_d$ each site will have exactly one electron. Consequently, the system will be a paramagnet with magnetic moments $\mu = \mu_B$, where $\mu_B$ is the Bohr magneton.

(b) It is not difficult to recognize that the "mean field" Hamiltonian is equivalent to the Hamiltonian investigated in Problem 5.2. Let us introduce $E_0 = E_d + U\langle n_{d,\downarrow}\rangle$. With the trial wavefunction, the Schrödinger equation becomes, for each $\boldsymbol{k}$,

$$\begin{pmatrix} E_c(\boldsymbol{k}) - E(\boldsymbol{k}) & \Delta \\ \Delta & E_0 - E(\boldsymbol{k}) \end{pmatrix} \begin{pmatrix} \alpha(\boldsymbol{k}) \\ \beta(\boldsymbol{k}) \end{pmatrix} = 0 \ . \tag{II.9.53}$$

The energies are

$$E = (E_c + E_0)/2 \pm \sqrt{(E_c - E_0)^2/4 + \Delta^2} \ . \tag{II.9.54}$$

The normalization condition gives a corresponding value to the weight factor $\beta$:

$$\beta^2 = \frac{\Delta^2}{(E - E_0)^2 + \Delta^2} . \tag{II.9.55}$$

The density of states was determined in Problem 4.10: $g(E) = g_c(E_c)(1 + \Delta^2/(E - E_0)^2)$. Remember, in a small range around $E_0$ the density of states is zero, in order to make the total number of states in the band correct. From this we obtain

$$N_d = \wp \int_{-\infty}^{\infty} g_c(E_c) \frac{\Delta^2}{(E - E_0)^2} \, dE, \tag{II.9.56}$$

where the symbol $\wp$ indicates the exclusion of the proper range around $E_0$.

Using the density of states, we obtain the spin occupation number:

$$
\begin{aligned}
\langle n_{d,\uparrow} \rangle &= \frac{1}{N_d} \int_{-\infty}^{E_F} g(E)\beta^2(E) \, dE \\
&= \wp \int_{-\infty}^{E_F} g_c(E_c) \left( 1 + \frac{\Delta^2}{(E - E_0)^2} \right) \frac{\Delta^2}{(E - E_0)^2 + \Delta^2} \, dE \\
&= \wp \int_{-\infty}^{E_F} g_c(E_c) \frac{\Delta^2}{(E - E_0)^2} \, dE .
\end{aligned} \tag{II.9.57}
$$

Note that this is exactly the same type of integrand as in Eq. II.9.56, except now the limit of integration is $E_F$. The second term in the integrand peaks around $E_0$; it is therefore reasonable to replace $g_c(E_c)$ by a constant, $g_0 = g_c(E_F)$. In principle, $g_c(E_c)$ is energy-dependent since $E_c = E_c(E)$, but the dependence is not strong if $E_d$, $E_F$, $E_d + U$ are all well within the band.

The occupation number $\langle n_{d,\uparrow} \rangle$ as a function of $E_F$ is plotted in Figure II.9.12. The functional form is $\sim -g_0 \Delta^2/N_d(E - E_0) = -\delta/(E - E_0)$, except for the range around $E_0$, where the function is constant. It is worthwhile to mention that the quantity $\delta/\Delta = g_0 \Delta/N_d$ is always small. If the number of states in the conduction band is equal to the number of states in the $d$ band, then $\delta/\Delta = \Delta/W$, where $W \gg \Delta$ is the bandwidth.

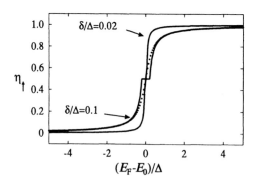

Fig. II.9.12. The occupation number for the $\uparrow$ spin state as a function of the Fermi energy for two values of $\delta/\Delta = g_0 \Delta/N_d$. The solid and dotted lines represent the exact and the approximate formula, respectively (for $\delta/\Delta = 0.02$ the difference between the two expression is not visible on the plot).

Up to this point our calculation was mathematically correct. However, the sharp kinks seen in Figure II.9.12 are not physical; they are the results of the simplifying assumptions made in the Hamiltonian Eq. I.9.25. Furthermore, the function does not seem to be very inviting to do further work with (although, in principle, we can continue our calculation by means of graphical solutions). Let us use a similar, smooth function:

$$\eta_\uparrow = \frac{1}{\pi} \arctan \frac{E - E_0}{\delta'} + \frac{1}{2}. \qquad (\text{II}.9.58)$$

with $E_0 = E_d + U\eta_\downarrow$. We also introduced the shorthand notations $\langle n_{d,\uparrow} \rangle = \eta_\uparrow$, $\langle n_{d,\downarrow} \rangle = \eta_\downarrow$. The parameter $\delta'$ must be selected such that the new function and the original one match reasonably well. Considering that the asymptotic behavior of $\arctan x$ is $1/x$, we find that the matching condition gives $\delta' = \pi\delta$. This smoother function is also plotted in Figure II.9.12 (dotted line). The smaller $\delta/\Delta$ is, the better the fit is.

The Hamiltonian for the $\downarrow$ spins gives a similar result, except the spins are interchanged. Self-consistency means that

$$\eta_\uparrow = \frac{1}{\pi} \arctan(\varepsilon - \gamma\eta_\downarrow) + \frac{1}{2} \qquad (\text{II}.9.59)$$

and

$$\eta_\downarrow = \frac{1}{\pi} \arctan(\varepsilon - \gamma\eta_\uparrow) + \frac{1}{2} \qquad (\text{II}.9.60)$$

are simultaneously satisfied [the symbols $\varepsilon = (E_F - E_d)/\pi\delta$ and $\gamma = U/\pi\delta$ were introduced]. For a representative range of parameters the graphical solutions for these equations are presented in Figure II.9.13.

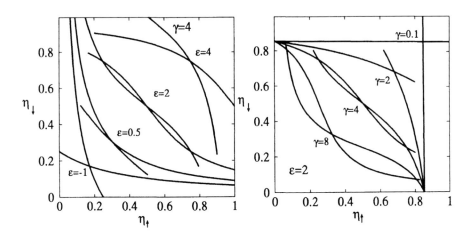

**Fig. II.9.13.** Self-consistency conditions at various values of the parameters $\varepsilon$ and $\gamma$.

If the the $d$ level lies slightly below the Fermi energy, then for small $U$ the only solutions have $\eta_\downarrow = \eta_\uparrow$; the system has no net magnetic moment on the $d$ sites (see Figure II.9.13, $\varepsilon = 2$, $\gamma = 0.1$ or 2). However, at a sufficiently large value of $U$ (like $\gamma = 4$ or 8 in the figure), two more solutions develop, with $\eta_\downarrow \neq \eta_\uparrow$. The energies belonging to these solutions are degenerate and lower than the energy of the $\eta_\downarrow = \eta_\uparrow$ solution. The system develops magnetic moments at the $d$ sites, but the magnitude of the moments are *not* equal to a moment corresponding to a fully localized spin: $\mu_{\text{eff}} = \mu_B(\eta_\downarrow - \eta_\uparrow) < \mu_B$.

For a fixed value of $U$ we see three different possibilities. If the $d$ level is above, or not too far below, the Fermi energy, the system is not magnetic with small and equal occupation of the two spin states on the $d$ levels (Figure II.9.13, $\varepsilon = -1$ or 0.5, $\gamma = 4$). When $E_d$ is well below $E_F$, the $d$ states are nearly full, but there is still no magnetism ($\varepsilon = 4$, $\gamma = 4$). However, for an intermediate situation ($\varepsilon = 2$, $\gamma = 4$), magnetization develops in the $d$ states. The situation resembles the $\Delta = 0$ case, discussed in part (a).

A short discussion of the basic ideas of the Anderson model can be found in the textbook by Harrison [5] p. 483. With the right set of parameters our calculation always results in a "broken symmetry" (i.e., magnetic) ground state. Also, the dimensionality of the system does not enter the calculation in any important way. We must emphasize that these features are due to the mean field approximation.

# References

1. N.W. Ashcroft and N.D. Mermin, *Solid State Physics*, W.B. Saunders, Philadelphia, 1976.
2. C. Kittel, *Introduction to Solid State Physics*, 7th edition, John Wiley & Sons, New York, 1996.
3. J.M. Ziman, *Principles of the Theory of Solids*, 2nd edition, Cambridge University Press, Cambridge, England, 1972.
4. H. Ibach and H. Lüth, *Solid State Physics, An Introduction to Theory and Experiment*, Springer-Verlag, Berlin, 1991.
5. W.A. Harrison, *Solid State Theory*, McGraw-Hill, New York, 1970.
6. P.Y Yu and M. Cardona, *Fundamentals of Semiconductors*, Springer-Verlag, Berlin, 1996.
7. R.A. Dunlap, *Experimental Physics, Modern Methods*, Oxford University Press, Oxford, England, 1988.
8. D.C. Mattis, *The Theory of Magnetism*, Harper & Row, New York, 1965.
9. L.D. Landau and E.M Lifshitz, *Electrodynamics of Continuous Media*, 2nd edition, Pergamon Press, Oxford, England, 1984.
10. H.B. Callen, *Thermodynamics*, John Wiley & Sons, New York, 1960.
11. N. Newbury et al. (eds.), *Princeton Problems in Physics*, Princeton University Press, Princeton, NJ, 1991.
12. A. Stella and L. Miglio (eds.) *Proc. of Int. School "Enrico Fermi": Semiconductor Superlattices and Interfaces*, North Holland, Amsterdam, 1993.
13. H.J. Goldsmid, *Problems in Solid State Physics*, Pion Limited, London, 1968.
14. J.M. Ziman, *Electrons and Phonons*, Oxford University Press, Oxford, England, 1960.
15. N.F. Mott and E.A. Davis, *Electronic Processes in Non-Crystalline Materials*, Clarendon Press, Oxford, England, 1979.
16. G. Grüner, *Density Waves in Solids*, Addison-Wesley, Reading, 1994.
17. F. Reif, *Fundamentals of Statistical and Thermal Physics*, McGraw-Hill, New York, 1965.
18. L.E. Reichl, *A Modern Course in Statistical Physics*, University of Texas Press, Austin, 1980.
19. See, for example, L.A. Woodward, *Introduction to the Theory of Molecular Vibrations and Vibrational Spectroscopy*, Oxford University Press, Oxford, England, 1972; or W.G. Fately, F.R. Dollish, N.T. McDevitt, and F.F. Beutley, *Infrared and Raman Selection Rules for Molecular and Lattice Vibrations*, John Wiley & Sons, New York, 1972.
20. M. Tinkham, *Introduction to Superconductivity*, McGraw-Hill, New York, 1975.
21. Landolt–Börnstein, New series, K.-H. Hellwege (ed.), Springer, New York, 1982.
22. N.P. Ong, *Phys. Rev. B* **43**, 193 (1991).
23. H.C. Montgomery, *J. Appl. Phys.*, **42**, 2971 (1971).
24. V.J. Emery, *Correlated Electron Systems*, World Scientific, Singapore, 1993.

# Index

255